Contested Grounds

SUNY Series in
International Environmental Policy and Theory

Sheldon Kamieniecki, editor

Contested Grounds

**Security and Conflict
in the
New Environmental Politics**

edited by
Daniel H. Deudney
and
Richard A. Matthew

State University of New York Press

Production by Ruth Fisher
Marketing by Dana E. Yanulavich

Published by
State University of New York Press, Albany

For information, address the State University of New York Press,
State University Plaza, Albany, NY 12246

Library of Congress Cataloging-in-Publication Data
Contested grounds : security and conflict in the new environmental
    politics / edited by Daniel H. Deudney and Richard A. Matthew.
        p.   cm. — (SUNY series in international environmental policy
    and theory)
        Includes bibliographical references and index.
        ISBN 0-7914-4115-6 (hc : alk. paper). — ISBN 0-7914-4116-4 (pb :
    alk. paper)
        1. Environmental policy.   2. Security, International—
    Environmental aspects.   3. Geopolitics.   I. Deudney, Daniel.
    II. Matthew, Richard Anthony.   III. Series.
    GE170.C6429   1999
    363.7'0526—dc21   98-43907
                                                                              CIP

10 9 8 7 6 5 4 3 2 1

Contents

Acknowledgments   vii

1. Introduction: Mapping Contested Grounds
   • Richard A. Matthew   1

Part I: Historical and Conceptual Background

2. Bringing Nature Back In: Geopolitical Theory from the
   Greeks to the Global Era • Daniel H. Deudney   25

Part II: The Contemporary Debate

3. Thresholds of Turmoil: Environmental Scarcities and
   Violent Conflict • Thomas F. Homer-Dixon   61

4. A Realist's Conceptual Definition of Environmental
   Security • Michel Frédérick   91

5. The Case for DOD Involvement in Environmental
   Security • Kent Hughes Butts   109

6. The Case for Comprehensive Security
   • Eric K. Stern   127

7. Threats from the South? Geopolitics, Equity, and
   Environmental Security • Simon Dalby   155

8. Environmental Security: A Critique
   • Daniel H. Deudney   187

Part III: Case Studies

9. Transboundary Resource Disputes and Their
   Resolution • Miriam R. Lowi   223

10. Imminent Political Conflicts Arising from China's
    Population Crisis • Jack A. Goldstone   247

11. Out of Focus: U.S. Military Satellites and
    Environmental Rescue • Ronald J. Deibert   267

Conclusion

12. Conclusion: Settling Contested Grounds
    • Richard A. Matthew   291

About the Contributors   303
Index   305

Acknowledgments

In May, 1993, most of the contributors to this volume met in Vancouver, Canada for a two-day conference. Papers were presented and discussed and the decision was made to consider producing an edited volume. The conference was made possible by grants from the Social Sciences and Humanities Research Council of Canada and the Institute for International Relations at the University of British Columbia, which also hosted the meeting. Several people deserve a special acknowledgment for their roles in organizing the conference: Ronald Deibert, Karen Gutierri, Brian Job, Richard Matthew, and Shannon Selin.

The task of transforming the conference papers into an edited volume was greatly assisted by the support of the Killam Foundation and the Canadian Institute for International Studies. In addition, the Focused Research Group on International Cooperation, the School of Social Ecology at the University of California at Irvine, and the Department of Political Science at the University of Pennsylvania provided invaluable institutional and financial support.

The chapters by Thomas Homer-Dixon, Jack A. Goldstone, and Miriam Lowi, were originally prepared for the Project on Environmental Change and Acute Conflict, directed by Thomas F. Homer-Dixon. The editors are grateful for permission to include revised versions of the project's working papers in this volume.

Finally, the editors would like to thank the following individuals and groups for encouragement, criticism, and advice: The Academic Council on the United Nations System, Ken Conca, Suzanne Cornwell, Geoffrey Dabelko, Mary Keeler, Kai Lee, Christa Sheehan, P. J. Simmons, Paul Wapner, and Mark Zacher.

1

Introduction:

Mapping Contested Grounds

Richard A. Matthew

When the environmentalist Lester Brown argued for a redefinition of national security in 1977, his work elicited little response among students of world politics. Six years later, Richard Ullman gave support to this initiative with a short article entitled "Redefining Security," in which he sought to broaden the concept of national security to include nonmilitary threats to a state's range of policy options or the quality of life of its citizens.[1] Brown and Ullman inspired a few scholars to reconsider the concept of security from an environmental perspective, but, during this period, the perceived imperatives of the Cold War continued to dominate both theory and practice in the area of security affairs.

The past several years, however, have seen a dramatic ground swell of interest in environmental change as a variable relevant to understanding security and conflict. This interest has not been confined to the academic world. In 1991, former President Bush added environmental issues to the "National Security Strategy of the United States." High level officials and academics now meet regularly to devise answers to the questions addressed in this volume.

The ground swell of interest has produced a number of important, although often controversial and inconclusive, empirical findings about environmental change as a source of insecurity and conflict. It has generated a lively exchange between those who view the redefinition of security as part of a general project to transform the international system, those who share this ambition but are skeptical of such an approach to realizing it, those who seek to incorporate environmentalism into existing institutions and practices, and those who regard the entire exercise as a passing fad related to the policy confusion and transitional unipolar moment that have followed the sudden end of the Cold War. Above all, this "second look" has contributed to the expanding field of environmental politics and broadened appreciation of the complexity of the environmental crisis.

This volume has been prepared to introduce students and practitioners to the theoretical debate and empirical evidence available today. In the following pages, I provide a general context for the subject by briefly presenting definitional moments in the history of the new environmental politics. I then sketch the early contours of the debates over environmental security and conflict. Next, I summarize the main questions explored in this volume. Finally, I review the contents of *Contested Grounds* on a chapter-by-chapter basis.

The New Environmental Politics

Although conservation movements, concerns about the deleterious impact of industrial pollution, and fears of scarcity-induced conflict and misery have received some attention throughout the industrial age, the emergence of environmental politics is a recent phenomenon.[2] It was during the turbulent decade of the 1960s that environmentalism began to assume its contemporary political form. Environmental activists, buttressed by scientific research, channeled mounting public anxiety about environmental change into a political movement that quickly began to affect political agendas at the local, national, regional, and global levels.

The anxiety was catalyzed and disseminated by a number of popular books, the most influential of which was Rachel Carson's controversial 1962 bestseller, *Silent Spring*. Carson's account of the impact of pesticides on human health and her moral outrage at the arrogance that permitted such behavior anticipated a revolutionary change in the manner in which the relationship between nature and civilization would henceforth be perceived. No longer could nature be regarded as simply raw material to be endlessly transformed by human ingenuity and labor into commodities. The relationship was more complex and delicate than previously suspected. Starkly put, the environmental life-support system upon which all life depended was being altered and degraded by human actions—at stake was the future of humankind.

By 1970, the groundwork was in place for Earth Day, "the largest environmental demonstration in history."[3] The social context that mobilized millions of Americans to participate in this event and supported the emergence of the new environmental politics has been described by John McCormick in terms of a general malaise about the broader implications and future of industrial affluence, the psychological stress of nuclearism, growing public alarm about environmental disasters, advances in scientific knowledge, and the compatibility of environmentalism with other antiestablishment programs such as the antiwar and women's movements.[4]

These and other themes were reflected in the foreboding literature that appeared at this time and underscored the global magnitude of the "environmental crisis." A vigorous debate erupted in public fora, nourished by the widely read works of writers such as Paul Ehrlich (1968), Garret Hardin (1968), Barry Commoner (1971), Donella Meadows et al. (1972), and Lester Brown (1972). By drawing attention, respectively, to issues such as exponential population growth, the "tragedy of the commons," the negative externalities of production technologies, the potential limits to industrial growth, and the complex global interdependencies of the late twentieth century, these authors provided the new environmental politics with a rich analytical and normative discourse that immediately engaged students and practitioners of world politics.[5]

Environmental issues were placed squarely on the agenda of world politics at the United Nations Conference on the Environment (1972) held in Stockholm. As Lynton Keith Caldwell notes, during the century prior to 1972, both governmental and nongovernmental members of the international community had met sporadically, and largely ineffectually, to discuss a range of environmental issues.[6] For example, the conservation and equitable distribution of resources was broached at the United Nations Scientific Conference on the Conservation and Utilization of Resources (1949). A number of recommendations related to research and education issued from the Intergovernmental Conference of Experts on a Scientific Basis for a Rational Use and Conservation of the Resources of the Biosphere (1968).

But it was at Stockholm that the international importance of environmental issues was clearly and officially recognized and given an institutional setting through the creation of the United Nations Environment Programme.[7] Moreover, the centrality of North-South issues and the vital role of nongovernmental organizations (NGOs) in the new environmental politics were both clearly acknowledged at the Stockholm conference.[8]

Building on the legacy of Stockholm, the past three decades have witnessed a flurry of activity at the international level. Over two hundred multilateral agreements have been negotiated addressing issues such as climate change, sea pollution, the use of nuclear materials, the protection of flora and fauna, air pollution, the military use of environmental modification techniques, and the transboundary movement of hazardous materials. Although many states have failed to sign these agreements and monitoring and enforcement remain imperfect, a corpus of international environmental law now exists to guide and regulate state actions. Indeed, international law scholar Dorothy Jones has argued that protection of the environment is emerging as a fundamental legal norm in the international system.[9]

Regional organizations as diverse as the Organization of African Unity, the European Union, the North Atlantic Treaty Organization, the Organization for Economic Cooperation and Development, and Asia Pacific Economic Cooperation have engaged in some level of environmental activity. The United Nations system, hampered by various organizational and political constraints, has acted to incorporate environmental issues into many of its specialized agencies, including the United Nations Development Programme, the World Bank, the Food and Agricultural Organization, the International Labour Organization, the World Health Organization, the International Maritime Organization, and the United Nations Education, Scientific, and Cultural Organization.

Throughout the 1970s and 1980s, public concern about the immediate and cumulative effects of environmental change, backed by increasingly sophisticated scientific research on problems such as acid precipitation and deforestation, compelled state officials to take environmental issues seriously. A major step forward occurred in 1983 when the United Nations General Assembly established the World Commission on Environment and Development. Chaired by the former Prime Minister of Norway, Gro Harlem Brundtland, the Commission released its report, *Our Common Future*, in 1987. Focusing on the global and interlocking processes of population growth, food production, ecosystem protection, energy use, industrialization, and urbanization, the report contained a wide range of proposals and recommendations woven together by the concept of sustainable development: development designed to "meet the needs of the present without compromising the ability of future generations to meet their own needs."[10]

The concept of sustainable development, negotiated in an attempt to bridge the diverse interests of developed and developing states, was elaborated at the 1992 Conference on Environment and Development held in Rio de Janeiro and has been integral to discussions at other U.N. conferences on issues such as population and development (Cairo 1994). Thus, over a twenty-five-year period, world environmental politics has evolved from the Stockholm generation's recognition of the seriousness of the problem to the Rio generation's attempt to design a solution to it—*Agenda 21*.[11]

Important developments also have transpired in the nongovernmental realm. NGOs such as Friends of the Earth (established 1969) and Greenpeace (established 1972) have become vital transnational forces, raising public awareness, engaging in political activism and scientific research, monitoring compliance with regimes, and participating in the NGO fora that take place alongside U.N. and other international conferences. Covering the entire political spectrum from reactionary to radical, ranging from highly specialized to broadly focused, and—depending on how one defines them—numbering in the tens of thousands, NGOs play a key role in shaping and supporting the new environmental politics of the late twentieth century.[12]

Through these various activities, three broad and interconnected issue areas gradually have emerged that today guide research, policymaking and activism: environmental ethics, sustainable development, and environmental security. These are clearly associated with the traditional concerns of students and practitioners of world politics: human rights and world justice; international political economy; and national security, war, and peace. In a field that became highly institutionalized during the Cold War era, it is not surprising that the predominant tendency has been to place environmental issues into familiar analytical and policy categories.

As a result, the conceptual issues of environmental politics have become increasingly integrated into the enduring and fundamental intellectual debates about realism, liberalism, and marxism; about structural and process explanations and prescriptions; about legal and market forms of regulation; about the utility of domestic and international institutions; and about the causes and consequences of conflict and cooperation, wealth and poverty, and justice and inequity.

While many are encouraged by the fact that environmental issues have moved into the political mainstream and gained legitimacy in the academic world, others fear that in doing so they have been diluted, losing their revolutionary potential and enabling scholars and policymakers to proffer compromised, short-term solutions designed to protect the status quo at a time when fundamental change is required. A common criticism leveled at the new environmental politics is that it has been coopted by the mainstream interests of Northern industrial states and now is governed by an agenda that marginalizes the concerns of the developing world while exaggerating its contribution to the environmental crisis. Proponents of this position, such as Maria Mies (1986), Vandana Shiva (1989), and Carolyn Merchant (1992), tend to endorse radical systemic change and frequently support grassroots movements and variants of deep ecology activism. Among environmental activists, similar concerns can be detected. Organizations such as Greenpeace have splintered as they have moved toward a more central and moderate position in the political arena, leaving some of their members frustrated by the perceived "sell-out."

Finally, a small minority of thinkers, such as Julian Simon and Herman Kahn (1984), Ronald Bailey (1995) and Wallace Kaufman (1994), have challenged the very utility of the new environmental politics on the grounds that its fundamental claim—that certain human activities affect the environment in adverse ways that threaten both the welfare of humankind and nature's complex evolutionary and recuperative processes—is misguided and alarmist. According to this critical perspective, environmental politics attracts resources away from productive enterprises in order to fatten already bloated bureaucracies and underwrite dubious academic ventures. This position's appeal relies heavily on the fact that scientific, demographic, economic, and political studies are often inconclusive in relating human actions to environmental change, and environmental change to threats to human welfare.

Because of the diversity of environmental concerns, the different and shifting priorities of various communities, and disagreements over causes, effects, and responses, it is not possible to describe in detail the many perspectives that together constitute contemporary environmental politics. The following typology, however, conveys a sense of this dynamic area.

Typology of the New Environmental Politics

Deep Ecology	Social Justice Ecology	Technological Optimism	Conservationism	Eco-Skepticism

Deep ecologists advocate a biocentric and holistic approach in which all forms of life are intrinsically valuable and interrelated. Technology, pride, and greed are among the forces that have encouraged *homo sapiens sapiens* to exploit, alter, and try to control the great web of life of which it is a part. In consequence, humans have become alienated from nature, isolated by layers of technology. Our activities have grown increasingly destructive, and we, along with many other species, are paying the price. We must try to reconnect with nature, rediscover its rhythms and patterns, and minimize our impact on it.

Social justice ecologists adopt an anthropocentric perspective. We must understand and address environmental change in terms of its relationships to forms of social injustice and violence, such as the oppression of women and indigenous peoples, the massive inequalities of wealth and power, increasingly destructive wars, and the systematic violations of human rights that characterize world politics. Day after day it is the marginalized and the poor who are in the front lines of global suffering—vulnerable to disease, compelled to move due to land degradation, and victims of scarcity. The exploitation and degradation of our environment and our fellow humans go hand in hand. Unless we adopt an approach to environmental rescue guided by principles of justice and morality, we may unwittingly implement policies that further divide humankind, and that sacrifice the many to maintain the privileged lifestyles of the few.

Technological optimists vary enormously in their assessments of environmental trends and conditions, but are united in their conviction that more efficient and environmentally sensitive technologies exist or can be developed that will resolve most, if not all, environmental problems. Their interest is in how to encourage the development, diffusion, and implementation of green technology. Sharing data, promoting technology transfer, and harnessing the tremendous power of the global market—supported by government regulation and consumer advocacy—are their preferred strategies.

Conservationists, who often root themselves in the emotionally and symbolically fertile ground of the nineteenth-century conservation groups that emerged in America and Europe, tend to be wary of "environmental alarmism." No population bomb has been detonated, few ecosystems have crashed, no real scarcity of environmental goods exists. While humans can be thoughtless and immoderate, common sense, informed by scientific research, encouraged by market incentives, and directed by prudent government regulation should guide our behavior.

Finally, the *eco-skeptics*, as previously noted, are openly hostile to the new environmental politics. The human species, some six billion strong and growing, is better off than at any time in its 200,000 year history. If the problems are real, our species will adapt as it has done in the past. It survived the peak of the last Ice Age, some 18,000 to 20,000 years ago with little more than animal skins and stone tools. It can survive a few degrees of global warming or the extinction of spotted owls.

Proponents of these various positions are evident in each of the three issue areas noted earlier. The concepts of environmental security and conflict have elicited a particularly vibrant debate among academics and policymakers alike that displays much of the diversity described above. The following section reviews the early contours of this debate and is very brief because many parts of this debate are summarized and discussed in subsequent chapters.

Environmental Security and Conflict: The Debate Unfolds

Simon and Kahn notwithstanding, after three decades of wide-ranging research and discussion, it is reasonable to assert that the expanding patterns of production, consumption, settlement, and waste disposal developed by the human species are adversely affecting the air, water, land, and biodiversity upon which all forms of life depend. Although some activists and intellectuals have endorsed an uncompromising ecocentric position, mainstream environmental politics is principally concerned with what this means for the welfare, security, and freedom of all or part of humankind.

The environmental security debate that has emerged within the new environmental politics has been shaped by two closely related clusters of questions. First, what does and should the concept of "environmental security" mean? Should the emphasis be on protecting the environment or addressing environmental threats to the security of states or of humankind? Is the impact of environmental change manifest in familiar forms of violence and conflict, new forms such as a gradual deterioration of the quality of life, or both, or neither? Do disagreements on the meaning of environmental security reflect deeper disagreements between the North and the South, men and women, elites and nonelites, or Western and non-Western cultures? In a world characterized by multiple forms of violence and innumerable sources of insecurity, where does environmental change rank?

Second, what are the risks involved in using a vocabulary that, in the arena of world politics, tends to evoke images of war and invite military participation? Are values such as peace and justice receiving adequate attention in this discourse? To what extent has it been fueled by post–Cold War concerns about cuts in defense spending? Can the military, with its vast resources, play a constructive role? How persuasive are the criticisms of those who fear that environmental politics is becoming a reactionary prop for entrenched interests instead of a revolutionary tool for change? Answers to these questions have evolved somewhat independently in the policy and academic communities.

Environmental issues have a fairly recent and marginal, but not insignificant, status in the security policy community.[13] In the 1970s, the OPEC oil crisis and "limits to growth" thesis stimulated concerns about how resource scarcity might jeopardize the economies of advanced industrial states and promote conflict. The concept of economic security emerged to address these concerns. Partially in response to this, the Carter Doctrine was announced, affirming the strategic value of the oil-rich Middle East. However, discussions of energy self-sufficiency as a national priority garnered little support. Throughout most of the 1980s, economic growth remained a domestic priority and security thinking focused on the Cold War rivalry with the Soviet Union.

The end of the Cold War created an opportunity to reconsider the concept of national security—and the potential threat posed by environmental change. Arguments developed within the policy community generally (1) underscore the immediate and prevalent nature of the threat; (2) relate it to national interests; (3) contend that existing beliefs, institutions, and practices are in some way inadequate; and (4) call for resources to be applied through new institutions or strategies to achieve specific objectives. The tone of these arguments is usually urgent and dramatic, designed to attract the support of officials concerned about the implications of institutional restructuring, and worried about climbing on a new bandwagon that might suddenly fall on its side.

The most articulate and influential arguments have been advanced by Jessica Tuchman Mathews. In her widely cited article "Redefining Security," Mathews endorses "broadening [the] definition of national security to include resource, environmental and demographic issues."[14] Pointing to the interrelated impact of population growth and resource scarcity, she imagines a bleak future of "[h]uman suffering and turmoil," conditions ripe for "authoritarian government," and "refugees . . . spreading the environmental stress that originally forced them from their homes."[15] Turning to the planetary problems of climate change and ozone depletion she completes a "grim sketch of conditions in 2050," and concludes with a set of general policy recommendations, entailing significant institutional change and aimed at ensuring this grim sketch does not become reality: slow population growth, encourage sustainable development, and promote multilateral cooperation.[16] More immediately, she argues, the United States should seek the elimination of ozone-depleting CFCs, support the Tropical Forestry Action Plan, support family planning programs, and develop a green energy policy.

A 1994 *Atlantic Monthly* article by Robert Kaplan, entitled "The Coming Anarchy," drew an even grimmer portrait of growing human misery, population displacement, violence, and conflict, related it to environmental degradation, and asserted that this was "the national security issue of the early twenty-first century."[17] This, too, has prompted discussion within the policy community, where a principal concern is to identify threats to U.S. interests and the image of chaos in the Third World appears rich with menacing possibility.

A more subdued but equally menacing article by Matthew Connelly and Paul Kennedy highlights the demographic dimension of the threat. "Must It Be the West Against the Rest?" draws heavily upon Jean Raspail's unsettling vision, in *The Camp of the Saints* (1973), of humankind's most miserable members leaving their bleak Third World existence in a desperate final effort to experience Western opulence. Connelly and Kennedy take this scenario seriously and offer a familiar set of policy recommendations if the West wishes to avoid it: give more aid and contraceptives, develop alternative energy sources, curtail arms sales, strengthen the U.N.'s interventionary capability, and promote a culturally sensitive code of human-rights-based ethics.

These writers, aware of current research on environmental change, sensitive to the sort of language that will attract policymakers, and building on themes that acquired legitimacy in the 1970s, have served as vital but selective conduits between the academic and policy worlds. Mathews's commitment to an interdependent and global conception of environmental security and her strong endorsement of multilateral solutions serve to coax policymakers away from conventional realist positions based on protecting explicitly national interests with strong military capabilities. Kaplan's impact is more difficult to assess. His penchant for sensationalism may prove to be galvanizing or destructive of environmental politics. Kennedy and Connelly remind us with great clarity of the need for a new North-South package of reforms.

In the United States, at least, policymakers appear to be listening. In recent years the Department of Defense has stepped up recycling and clean-up efforts, cooperated with foreign militaries to develop environmental guidelines, worked with other agencies at home and abroad to address problems such as radioactive contamination in the Arctic, and hosted regional conferences on environmental conflict and security. The intelligence community has begun to track and analyze environmental change, opened some of its archived satellite imagery to scientists, and promised to make its extensive data collection and assessment assets available to support environmental policy. The State Department has vowed to green American foreign policy by publishing annual reports on its goals and accomplishments, spearheading international efforts to improve treaty monitoring and enforcement, and establishing regional opportunity hubs in strategic locations in order to integrate environmental concerns into its operations and serve as sites for addressing pressing environmental problems. The general attitude informing these and many other initiatives is that the interests of the United States and the world depend on a healthy environment. Clearly, the concept of environmental security has been an important catalyst to these efforts, although assessments of their impact and potential vary considerably.[18]

While the writers noted above may be having the greatest immediate impact on policy, it is in the academic world that the concept of environmental security has been explored in depth and from a wide range of perspectives that will have a long-term influence on policy.

While some scholars such as Gray and Rivkin (1991) have expressed skepticism about any relationship between environmental change and security, most of those who study this issue agree that environmental change threatens human welfare in some way. There is sharp disagreement, however, on how best to apply which resources to what ends. These disagreements reflect different levels of analysis, different interpretations of empirical evidence and causal chains, and different normative biases.

In large measure, these disagreements can be traced to the long-standing divide in world politics between those who seek to protect and refine a liberal world order of sovereign nation-states, markets, and regimes and those who seek to transform the current international system on the grounds that its states, markets, and regimes embody fundamentally unjust or undesirable values and practices. The former relate environmental security to the preservation of the international system; the latter to its radical transformation. Thus one dimension of the debate has been shaped by the confrontation between statists and globalists, reformists and radicals, liberals and their critics. While both sides agree that existing economic and political practices have caused the current environmental crisis, they part on the question of whether these practices need to be revised or replaced.

This is a difficult question to answer. Many scholars and policy-makers agree in some measure with Francis Fukuyama's bold claim that the big questions have been resolved. Humankind has determined the ideal forms of politics (representative democracy), economics (market economies), and ethics (human rights). On the ground, however, these systems are often inefficient and imperfect. Our challenge now is to use technology and other tools to solve problems within these "ideal" systems.[19]

Many others, however, contend that our problems are of a more fundamental nature. Either the core values (such as sustained economic growth) that guide our social structures are misguided and need to be changed (a reformist view), or the structures themselves are undesirable and need to be replaced (a revolutionary view).

This fundamental divide shapes much of the debate, but it is not the sole primary source of disagreement. Another fundamental—and crosscutting—divide is evident, although often cloaked in the shadows of academic discourse, in two markedly different images of what environmental security requires. Here a powerful technocratic-managerial image—inevitably biased toward the technical skills of the North—competes with an equally powerful, but less widely endorsed, democratic image. Thus, at the most general level, the debate over environmental security ranges from a position advocating the preservation of the status quo through the management of Northern elites to change inspired and governed by a global democratic politics. Between these extremes lie conceptions of the preservation of the status quo through some form of democratization and change guided and managed by transnational elites.

A third complication stems from the fact that the consequences of environmental degradation are experienced differently over time and space. From the perspective of someone living in a poor, sub-Saharan state, a sense of immediate threat may result from water pollution and scarcity, soil erosion, or the spread of disease. A Western European, on the other hand, may feel more threatened by the long-term effects of radioactive contamination or global warming. For the former, deforestation may be seen as essential to survival, whereas for the latter it may be seen as a source of climate change or gene pool loss that should be halted. Consequently, the policy agendas that emerge to ensure environmental security vary enormously, making multilateral agreement and cooperation difficult. Coordination problems increase dramatically when material interests, future discount rates, and the benefits of competitive extraction vary.

In light of these differences, it is only at a very high level of generality that one can speak of environmental security as a clear and distinct concept appropriate to the entire world. On these terms, environmental security might be defined as a condition having three characteristics: First, it is a condition in which environmental goods —such as water, air, energy, and fisheries—are exploited at a sustainable rate. Second, it is a condition in which fair and reliable access to environmental goods is universal. And third, it is a condition in which institutions are competent to address the inevitable crises and manage the likely conflicts associated with different forms of scarcity and degradation.

Can the systems in place today provide environmental security on these terms? And, if so, will this be achieved through representative or participatory strategies? In any case, such a concept is impossible to operationalize fully in the foreseeable future. We inhabit a world in which environmental change affects human welfare and freedom in diverse ways; in which huge inequalities of wealth and power, combined with the random distribution of environmental goods, create very different payoff structures for international actors; in which existing institutions are only beginning to take environmental factors into consideration; and in which the environmental agenda is itself subjected to the pull of other human agendas.

Thus, while it is possible to define environmental security in an idealized manner appropriate to the entire planet, particular understandings that are often at odds with each other litter and shape the contemporary landscape. At the very least, it is possible to identify four distinct conceptualizations:

1. *Deep Ecology.* As discussed earlier, this perspective stresses

the security of our entire interactive and interdependent planetary environment.

2. *Human Security.* This position is captured in the social justice definition offered above. Its emphasis is on ensuring that all of humankind has fair and reasonable access to a healthy environment, today and tomorrow.

3. *National Environmental Security.* This is the position that is most evident in the environmental security policy that has developed in the United States in recent years.[20] Its focus is on (a) greening military training, testing, and warfighting activities, (b) using military and intelligence assets (such as data collection and analysis capabilities) to support environmental policy, (c) tracking environmental problems that might trigger, generate, or amplify violent conflict, (d) developing anticipatory policies for dealing with environmentally stressed areas, and (e) integrating environmental concerns into conflict resolution processes.

4. *Rejectionist.* The rejectionist position is not a variant of eco-skepticism; rather, its advocates contend that linking environmentalism and security is not a desirable approach because environmental change rarely causes a conventional security problem and security instruments are poorly designed to deal with environmental problems.

In short, different perspectives on fundamental structures, operational strategies, and environmental priorities thicken an idealized formulation of environmental security in at least four distinctive ways. Since the clearest examples of these various positions are presented in subsequent chapters, it is appropriate to allow the reader to examine the debates first hand and to turn to the central questions unifying the chapters of this volume.

The Main Questions

Three main questions have been raised and addressed in this volume:

1. What are the various meanings ascribed to the concept of environmental security today, how significant are the differ-

ences, and what are the risks involved in accepting and building upon this term? Several answers to this question have been outlined above; these and others are developed more fully in subsequent chapters.

2. What is the relationship between environmental change and conflict or other forms of violence? In responding to this question, contributors to this volume have sought not only to clarify the relationship but also to gauge its significance as a security threat and to consider the possibility that environmental change might, at least in some cases, be better characterized as a motivation for cooperation.

3. Can the discourse of environmental security be harnessed to the formulation and implementation of effective research agendas and policies? Of particular significance is the current tension between defense conversion advocates and proponents of enlightened military strategy. A less explicit but perhaps more important tension exists between managerial—especially Northern and technocratic managerial—policy responses and more democratic and global initiatives. Beneath these policy preferences lie competing agendas for preserving the status quo and promoting fundamental change.

These are not the only questions addressed in this volume and the authors were not asked to respond to them directly. Rather, these are the main questions that, in retrospect, provide continuity to the various chapters and underlie many of the disagreements evident between them.

Chapter-by-Chapter Review

Following this introductory chapter, *Contested Grounds* is composed of three parts. Part I, "Historical and Conceptual Background," consists of a single chapter by Daniel Deudney entitled "Bringing Nature Back In: Geopolitical Theory from the Greeks to the Global Era." Deudney's analysis suggests that contemporary debates over environmental security can be enriched by two much older traditions of thought—one focused on nature as a cause of political outcomes, the other exploring conflict and cooperation from a geopolitical perspective. By describing the process through which these earlier traditions were modified and marginalized in the industrial era and have now resurfaced as innovations, Deudney hopes to recover insights that may be fruitful in understanding the rift between North and South.

Deudney carefully situates the concept of environmental security in a broader theoretical context to demonstrate its continuing historical relevance while still appreciating the particular challenges posed in the twentieth century that have given it its recent prominence. His analysis points out that (a) the security issues of our century have been generated in large measure by the various ways in which human and ecological systems have interacted over time; (b) the realist tradition, which appears least interested in grappling with environmental issues, was in fact constructed largely in response to this very problematic; and (c) weaknesses in contemporary international relations theory can in part be traced to the omission of natural variables, an omission that the debate on environmental security is correcting. Two of the case studies included in this volume (by Lowi and by Goldstone) are contemporary extensions of the more traditional problematics discussed by Deudney.

Part II, "The Contemporary Debate," includes seven chapters by scholars closely associated with recent debates over environmental security and conflict. This section of the volume covers a wide range of perspectives, and touches upon many of the key issues: the relationship of environmental change to conflict; the concern that environmental security is becoming a defense of the status quo; the realist position that much of the debate really addresses a very conventional concern; and the position of the defense establishment that, new or traditional, a military solution is feasible.

In "Thresholds of Turmoil: Environmental Scarcities and Violent Conflict," Thomas Homer-Dixon argues that violent conflicts throughout the developing world are being caused or exacerbated by resource scarcities. Reviewing the results of eight case studies conducted for the Project on Environmental Change and Acute Conflict (two of which are included in this volume), as well as evidence from other sources, Homer-Dixon suggests that this form of conflict is likely to increase as the pressures of environmental change overwhelm the capacity of institutions to adjust and respond, creating conditions for fragmentation or authoritarian government. This chapter demonstrates how the transnational phenomenon of environmental change may be related to the post–Cold War phenomenon of intrastate conflict, which is now the most common form of violence, while acknowledging the potential for this to escalate to the interstate level.

Homer-Dixon's work has been extremely influential in both the academic and policy communities. This chapter, which draws upon earlier published work, provides an updated statement of his position that takes into consideration concerns raised by other writers.[21]

An uncompromising state-centric approach to international relations is unpopular among many environmentalists but continues to dominate much of the theory and practice of the field. In "A Realist's Conceptual Definition of Environmental Security," Michel Frédérick advances a simple and concise, realist position. He defines environmental security as the "absence of non-conventional threats against the environmental substratum essential to the well-being of [a state's] population and to the maintenance of its functional integrity." Frédérick defends this state-centric perspective by underscoring the continuing centrality of the state in world politics, and the greater capacity of state institutions to act effectively in comparison to international organizations. Moreover, he argues, while stressing the importance of the state, his definition clearly distinguishes between threats that require a military response and those that do not, and thus creates a viable basis for cooperative strategies.

In "The Case for DOD Involvement in Environmental Security," Kent Butts argues for involving military institutions in the process of maximizing environmental security. As Butts notes, environmental security is already a part of the mission of the U.S. military. The question, then, is whether this role should be expanded or reduced and in what ways. Pointing to recent efforts by the military to change its status as a major polluter, and responding to many of the concerns raised by Deudney and others, Butts argues that the military has extensive resources and skills that can be applied effectively to both domestic and international environmental security issues without compromising its war-fighting capabilities. Moreover, the U.S. military has the potential to influence military establishments in other countries in ways beneficial to U.S. interests, global security, and the environment.

In "The Case for Comprehensive Security," Eric Stern argues for a comprehensive and multifaceted concept of security that includes an environmental component. In this way Stern hopes to address the concern expressed by Deudney, who opposes using the language of security, and Frédérick, who seeks to divide the concept of environmental security into traditional security problems and nonsecurity problems. Stern concludes that the concept of comprehensive security, which incorporates military, environmental, economic, political, and social values, provides the most constructive framework for effective, forward-looking policymaking.

Simon Dalby provides an important complement to the work of Homer-Dixon and Deudney in "Threats from the South? Geopolitics, Equity, and Environmental Security." In this chapter, Dalby examines the concept of environmental security in terms of both differences in the interests, experiences, and roles of Northern and Southern states, and tensions between managerial, status quo–oriented approaches, and more equitable, reform-oriented strategies for addressing the environmental crisis. By examining several environmental security issues from the perspectives of North and South, Dalby discloses disturbing trends in the evolution of the concept. As Dalby demonstrates, at stake is whether the concept will be employed to sustain traditional geopolitical understandings of security that favor the developed states or used to promote the protection of the global environment and all of its inhabitants.

Finally, Daniel Deudney has substantially revised earlier work calling into question both the utility of the concept of environmental security and the claim that environmental change tends to generate conflict.[22] In "Environmental Security: A Critique," he reiterates and expands upon three concerns. First, Deudney argues that environmental problems are conceptually unlike traditional security problems that focus on external aggression. While it is true that national security and environmentalism are linked insofar as military practices consume resources that could be applied to environmental rescue and often generate pollution, environmental degradation is unique in terms of the types of threat it poses, the sources of these threats, the extent to which they are intentional, and the kinds of institutions that are best suited to dealing with them. Second, it is dangerous, Deudney maintains, to try to harness the rhetorical and emotional allure of national security to environmentalism. The former is achieved through appeals to urgency, zero-sum thinking, and a "we versus they" mentality. In contrast, environmental change is a gradual and long-term threat that can best be addressed by building a sense of global solidarity based on shared interests. Finally, the language of security implies the likelihood of interstate conflict. But environmental threat is not likely to manifest itself in this way—the gradual immiseration of people is a more likely scenario. Deudney concludes that environmental change is best viewed as a global problem that challenges conceptions of national security. Instead of trying to adapt the latter, we should act to move beyond it and forge conceptions of security in international terms that better reflect the nature of the problem.

The last section of this volume, Part III, "Case Studies," contains three chapters that explore the issues raised above through focused case analysis. These three case studies include examples from both the developed and developing worlds, and cover resource scarcity and conflict, demographic issues, and the role of the military.

The first chapter serves to bridge traditional geopolitical issues with more recent environmental security debates by focusing on the problem of water scarcity. In "Transboundary Resource Disputes and Their Resolution," Miriam Lowi presents a detailed analysis of the complex nature of disputes over Jordan waters and their role in the Middle East peace process. Lowi argues that attempts by the United States, guided by functionalist theory, to resolve the conflict over water as a step toward a more general settlement were unsuccessful. Decoupling economic issues from political ones is not possible, she concludes, if the latter are characterized by deeply entrenched conflict. Moreover, while resolving political conflict may create conditions for developing a cooperative solution to the problem of water scarcity, the need for changes in consumption practices will not be easily addressed. Lowi's study has important implications for addressing cases in which resource scarcity is one of several sources of conflict.

The impact of population pressures on political stability in China is explored by Jack Goldstone in "Imminent Political Conflicts Arising from China's Population Crisis." Supporting many of the claims made by Homer-Dixon, Goldstone argues that the combination of population growth and over-burdened arable land has been a source of conflict in China for several hundred years. Goldstone contends that recent divisions within the ruling party and among elites, together with mounting difficulties in controlling Chinese society and appeasing discontented peasants and workers, have made the current regime extremely vulnerable. Add to this the fact that it will be very difficult to accommodate the needs of the tens of millions of Chinese who will be born in the next few decades, and the future appears bleak. Goldstone concludes that "[i]t seems unlikely that the collapse of communist China can be averted."

In the final chapter of this section, the role of the military in environmental security is examined by Ronald Deibert in "Out of Focus: U.S. Military Satellites and Environmental Rescue." Focusing on the possibility of using U.S. military satellites to support environmental protection and rescue projects, Deibert raises a number of concerns that reinforce and extend arguments made by Deudney and Dalby. By comparing military and civilian satellite systems, questioning the utility of declassifying military imagery, underscoring the military penchant for secrecy, and showing how during the Gulf War the military was able and willing to take over and censor civilian imagery, Deibert makes a forceful case for discouraging military involvement and encouraging the development of civilian capabilities.

Conclusion

The chapters in this volume do not resolve the various debates surrounding the concepts of environmental security and conflict. They do, however, provide a clear map of the areas of consensus, the principal disagreements, the conclusions of recent empirical studies, and the concerns that need to be addressed in the years ahead. Maintaining the environmental integrity of the planet and the welfare of humankind requires tough choices about both resources and institutions. There is no one path toward an environmentally secure future; there are many routes to conflict, violence, and misery. Avoiding the latter will demand innovation, pragmatism, and sacrifice. Students and practitioners of world politics must weigh different arguments carefully and act quickly and decisively in an era marked by skepticism and uncertainty, while remaining open to new ideas and information.

NOTES

1. Richard Ullman, "Redefining Security," *International Security*, vol. 8, p. 133.

2. For useful discussions of the nineteenth- and early twentieth-century background to contemporary environmental politics see, among others, Lynton Keith Caldwell, *International Environmental Policy: Emergence and Dimensions*, 2d ed. (Durham: Duke University Press, 1990); John McCormick, *Reclaiming Paradise: The Global Environmental Movement* (Bloomington: Indiana University Press, 1989); and Robert C. Paehlke, *Environmentalism and the Future of Progressive Politics* (New Haven, Conn.: Yale University Press, 1989).

3. John McCormick, *Reclaiming Paradise: The Global Environmental Movement* (Bloomington: Indiana University Press, 1989), p. 47.

4. John McCormick, *Reclaiming Paradise*, op. cit., pp. 49–64. Similar forces have been at work in many other countries of the world, although the

public response has varied considerably. In Western Europe, for example, where political behavior is often directed by parties, a number of Green parties have emerged and fought elections with mixed results.

5. Other writers focused on the domestic implications of environmental change, leading to a more explicitly domestic form of environmental politics that I do not discuss here.

6. Lynton Keith Caldwell, *International Environmental Policy: Emergence and Dimensions*, 2d ed. (Durham, N.C.: Duke University Press, 1990), pp. 30–54.

7. Lynton Keith Caldwell, *International Environmental Policy*, op. cit., p. 71.

8. John McCormick, *Reclaiming Paradise*, op. cit., p. 105.

9. Dorothy V. Jones, *Code of Peace: Ethics in the World of the Warlord States* (Chicago: University of Chicago Press, 1990), p. ix.

10. World Commission on Environment and Development, *Our Common Future* (Oxford: Oxford University Press, 1987), p. 8.

11. Unfortunately, as the five-year assessment of the Rio Plan and UNEP's first *Global Environmental Outlook Report* demonstrate, we are not acting quickly enough to stop, let alone reverse, the many forms of significant human-generated environmental change.

12. For a discussion of the role of NGOs, see Paul Wapner, *Environmental Activism and World Politics* (Albany: State University of New York Press, 1996).

13. This claim is based on recent personal experience in Washington with policymakers concerned with environmental security. It reflects their perception that they still have to work hard to introduce environmental concerns into the policy process, rather than the conclusions of an empirical study of this process.

14. Jessica T. Mathews, "Redefining Security," *Foreign Affairs* vol. 68, p. 162.

15. Jessica T. Mathews, "Redefining Security," op. cit., p. 168.

16. Jessica T. Mathews, "Redefining Security," op. cit., p. 172.

17. Robert D. Kaplan, "The Coming Anarchy," *The Atlantic Monthly* vol. 273 (1994), p. 45.

18. See Richard A. Matthew, "The Greening of U.S. Foreign Policy," *Issues in Science and Technology* vol. 13, no. 1 (Fall 1996), pp. 39–47.

19. See Francis Fukuyama, "The End of History?" in Foreign Affairs

Agenda, *The New Shape of World Politics* (New York: W. W. Norton, 1997), pp. 1–25.

20. Richard A. Matthew, "The Greening of U.S. Foreign Policy," op. cit.

21. Thomas Homer-Dixon, "Environmental Scarcities and Violent Conflict: Evidence from Cases," *International Security* vol. 19 (1994), pp. 5–40.

22. Daniel Deudney, "The Case Against Linking Environmental Degradation and National Security," *Millennium* vol. 19 (1990), p. 470.

PART I

Historical and
Conceptual Background

Bringing Nature Back In

Geopolitical Theory from the Greeks to the Global Era

Daniel H. Deudney

Recent years have witnessed an explosion of interest in the relationships between the physical environment and human affairs. One particularly active area of research has been "environmental security and conflict." The new concern for nature as a factor in human events marks a sharp departure from the main direction of post–World War II international-relations theory, which (with a few notable exceptions)[1] has neglected nature and sought to locate the social causes of social events.

The recent literature typically casts the natural environment as a new factor in politics, but in fact the idea that nature is a powerful force shaping human political institutions is extremely old. Arguments about nature as a cause of political outcomes were among the first in Western political science.[2] Analysis of nature as a cause was central to Aristotle and Montesquieu, the two major preindustrial writers universally identified as predecessors by twentieth-century political scientists (as opposed to political philosophers). Long before the "greenhouse effect" had even been discovered, let alone connected to human politics, Montesquieu had declared the "empire of climate is the first and greatest empire."[3] Long before the terms realpolitik or geopolitics were coined, political scientists had sought to understand nature as a cause of political events. Thus "bringing nature back in" returns political science to its earliest and most basic theories.

The new literature on environmental security and conflict is also generally antirealist in its rhetoric and content. At the same time, neorealists have sought to downplay environmental issues and exclude them from the "security studies" subfield of international relations.[4] Yet the recent environmental conflict and security literature emphasizes *conflict* as an outcome of natural forces, which is consistent with realism's assumption that conflict, scarcity, and insecurity are endemic in world politics. Thus, despite its antirealist rhetoric, the emergent environmental security paradigm has a great deal in common with realism. The sources of this peculiar estrangement between contemporary realists and analysts of environmental security are to be found in the narrowness of recent realist theory, and the historical shallowness and conceptual underdevelopment of environmental security theory.

The version of realism closest to environmental security concerns is not, however, the neorealism now dominant in American international relations theory, but rather an older and now neglected naturalist and geopolitical realism. This early naturalist and geopolitical theory is also the historical predecessor of contemporary environmental security studies. Unfortunately, the term *geopolitics* has lost the theoretical weight it once had, and has come to be used as a loose synonym for interstate power politics or to evoke an extremely conflictual view of international politics. The actual center of gravity of early naturalist, materialist, and environmental theory was the broad claim that variations in nature caused significant variations in human culture, society, economy, and politics.[5] The conventional association between geopolitics and conflict also conveys an incomplete and misleading view of the scope of geopolitical theory because geopolitics has traditionally concerned itself with the ways in which environmental factors are the cause of political order, cooperation, and interstate peace, as well as competition and conflict.

This chapter attempts to improve the conceptual sophistication and historical depth of the contemporary debates about environmental security by "bringing nature back in" in three ways. First, I examine several major conceptual issues involved in theorizing about nature as a cause of political outcomes. Simply inserting environmental factors into existing social scientific models is more problematic than recent researchers have recognized. Before nature can be brought back into social science, a set of thorny conceptual problems must be addressed. Because modern social science began its distinctive intellectual career by rejecting natural causes of social outcomes, the implications of the expulsion of nature from social science must be clarified and rethought.

Second, I aim to add historical depth to the environmental security and conflict debate by briefly describing and evaluating three early geopolitical arguments about climate, arable land, and Darwinian "survival of the fittest" metaphors. An examination of these arguments will demonstrate the range and complexity of early theorizing about natural sources of conflict, and the existence of an equally robust body of naturalist arguments about the environmental causes of cooperation and order.

Third, I provide a synoptic summary of arguments, both old and new, about the role of environmental factors in causing the global rift between the wealthy countries of the "North" and the poorer countries of the "South." Melding the arguments of early geopolitical writers with recent work by historians operating with implicitly naturalist approaches, the third section of the chapter illustrates the power of naturalist social science in explaining major features of world politics while challenging the currently dominant social scientific explanations for Third World underdevelopment.

Conceptualizing Nature as a Cause

Many of the propositions being advanced about the political consequences of the environmental change are causal chains that run from human-nature interactions, and/or nature-technology interactions to political outcomes. Neglected earlier naturalist and geopolitical theories explored such causal chains, and set forth a diverse array of claims about the natural environment as a cause of political, economic, and social outcomes. What does it mean to posit nature as a cause of human behavior? How do such theories differ from more conventional social science? How can naturalist theory avoid eliminating human agency and succumbing to complete determinism?

The Varieties of Naturalism

Naturalist theories are diverse, but all posit nature as a cause of human events. They disagree in their characterizations of nature, in the ways they claim nature shapes humanity, and in the human phenomenon they hold to be caused by natural forces. Naturalist thought comes in at least four[6] main varieties: *cosmic naturalism, analogical naturalism, anthropological naturalism,* and *functional-materialist naturalism.* The first two varieties of naturalism are pre-scientific or quasi-scientific, while the latter two types are more compatible with scientific understandings of human affairs.

First in breadth and antiquity, is *cosmic naturalism,* where an intuitively grasped cosmology is understood to be guiding human events. For example, astrologists locate the cause of human events in astronomical patterns, but do so unscientifically because propositions are not tested through empirical investigation, and the mechanisms linking causes and consequences are mysterious. Despite the cultural hegemony of science, recent concern over ecological decay has evoked a resurgence and reformulation of cosmic naturalist claims with concepts drawn from modern natural science, most notably mythopoetic formulations of the "Gaia Hypothesis."[7]

A second body of theoretical writings in which nonhuman physical nature plays a central role are *analogic naturalisms,* which hold that the logic of human social systems can be understood as analogous to the operation of natural ones. Although steeped in the images and languages of nature, such analogic naturalisms are not, strictly speaking, natural *theory* because they typically do not claim that nonhuman physical nature is the *cause* of human outcomes, but only that human social processes *parallel* natural ones. Many social science theories indulge in casual or episodic analogic naturalism by borrowing various metaphors and models from natural science, but this borrowing has been mutual, as when Charles Darwin applied "the doctrine of Thomas Malthus . . . to the whole animal and vegetable kingdoms."[8] By far the most elaborate and influential analogic naturalism is to be found in various versions of "social Darwinism."

The use of organic metaphors and analogies to describe politics is fraught with pitfalls and full of undisclosed and unexamined agendas. In general, there is no convincing reason to believe that deeply recurrent patterns in nonhuman nature—even if properly understood—must serve as a model for human interaction, because human beings differ in so many important ways from other animal and all plant life. Nor is it clear which part or process of nonhuman biological nature is analogous to which part or process of human social and political life.

Third, *anthropological naturalist* theories locate the causes of human behavior in intrinsic human nature—factors such as "race," intelligence, genetic endowment, and deep psychology.[9] Such theories, with a few exceptions[10] have largely disappeared from the study of international relations and political science, but they have not disappeared from the social sciences more generally, as evidenced by the continued vitality of sociobiology, physical anthropology, and biological psychology.

Fourth, of most relevance to issues of environmental security and conflict, are *functional-materialist naturalist* theories. Such theories are behavioral social scientific propositions about how specific human behaviors and institutions result from the interaction of humans with the nonhuman natural environment. These theories rest on the simple assumption that the physical world is not completely or even primarily subject to effective human control, and that these material contexts impede or enable vital and recurring human goals. Such theorists attempt to link specific physical constraints and opportunities given by nature (such as the presence of fertile soil, salubrious weather, access to the sea, and mountain ranges, etc.), to alterations in the performance of very basic functional tasks universal to human groups (most notably protection from violence and biological sustenance). Because humans conceive and carry out their projects in differing material environments, the various ways in which these environments present themselves to humans heavily shapes the viability of various human projects. Such theories are not exclusively about political outcomes, but encompass sociological, economic, and cultural[11] outcomes as well. The functional dimension of such theories stems from the positing of minimalist anthropological naturalist assumptions about naturally given human needs and capacities. Social activity is then assessed as being functionally adaptive or maladaptive in specific material contexts. Social activity is thus understood as mediations between a set of naturally given human needs and capacities, and the constraints and opportunities afforded by natural-material contexts.[12]

Geopolitics, or in more precise and less freighted terminology, *physiopolitics* (from *physis*, Greek for nature), is the subset of functional-materialist theories focusing on variations in the nonhuman physical environment that shape human *political* outcomes. Thus defined, this tradition of theorizing is as old as political science itself, and encompasses a vast theoretical literature stretching from Aristotle, through Montesquieu to the global geopoliticans of the late nineteenth century. The main effort of these early theories is to show that a handful of natural independent variables (most notably *climate, arable land, mineral resources, population,* and *topography*) shaped patterns of human political life in important and recurrent ways. The range of the dependent variables is also wide, encompassing military vulnerability and capability, patterns of political identity and authority, political competition and cooperation, and modes of production.

Denatured Social Science

What is the relationship between the geopolitical or physiopolitical branch of naturalist theory and social science? Some naturalist theories are social scientific, in that they attempt to explain repeated patterns of human life. But not all, or even most social scientific theories are naturalist or physiopolitical. Most social science attempts to explain human outcomes as the result of human social causes: behavioral outcomes are the result of processes and patterns of human sociality and group dynamics. Thus most social scientists seek social causes of social outcomes. In contrast, naturalist social scientists seek natural causes of social outcomes. Thus social science (the scientific study of human outcomes) has two branches, *natural-social science* (one branch of which is physiopolitics), and what for simplicity and clarity can be called *social-social science* (the study of the social causes of social effects.) Because of the decline of natural-social science most contemporary social science is *denatured* social science. The new concern for environmental decay requires a move beyond social-social science, and back to natural-social science.

The decline of natural-social science and the rise of social-social science were intimately connected. The emergence of social science as a separate branch of academic inquiry in the late nineteenth and early twentieth century entailed an often explicit break with the naturalism of previous political scientists. As the sociologist Samuel Klausner has observed, "in part, social science has asserted its independence as a discipline by demonstrating the limited explanatory power of physical concepts of human behavior."[13]

The tendency of social scientists to neglect natural variables in favor of human institutional and historical variables has intensified since World War II. In the new academic study of "international relations" defined as a branch of political science, the analysis of natural variables was largely absent. In part this resulted from the general discrediting of geopolitical theorizing produced by the links between Nazi aggression and the German geopolitical theory. The loss of earlier naturalist and materialist arguments is particularly visible in Kenneth Waltz's influential "three images" typology of theories of the causes of war.[14] Nature appears in the "first image" theories about human nature, but in Waltz's analysis of the "second image" domestic theories, and "third image" interstate theories, the earlier geopolitical lines of argument have nearly disappeared.

The expulsion of nature from social science in general, and political science in particular, has paralleled an opposite tendency in historical scholarship. The discipline of history exhibits diverse tendencies, but beginning in the late nineteenth century and widening in the twentieth, "history from below" has come to supplement, if not fully displace, the actor-centered history as the narrative of leaders, great events, and peoples. As the early-twentieth-century Berkeley social theorist Frederick Teggart observed, the older history "is the narrative statement of happenings which concern the fortunes and the existence of a particular nation." Such history "resembles closely that exemplified in Greek tragedy. They have described great and serious occurrences in the light of their outcome, and have sought to make the deeds of heroes and great men intelligible by the imaginative reconstruction of character."[15]

The turn to materialist analysis has been particularly pronounced among historians examining wide ranges of space and time. As the historian Geoffrey Barraclough observed, "the prevalent tendency [in world histories] is to adopt a broadly materialist position, in the sense that their central theme is man's conflict with his environment."[16] Without neglecting nonmaterial factors, historians such as Fernand Braudel, William McNeill, Alfred Crosby, and Geoffrey Barraclough have relied heavily upon material variables in attempting to explain patterns of regional and global history. Thus Braudel, reversing Leopold von Ranke's definition of the historian's enterprise, describes the mountain as more important than the ruler. Braudelian "structures" of the *long duree* are the largely static material realities of "man in his intimate relationship to the earth which bears and feeds him."[17] This rich and growing body of materialist history fits easily with the conceptual frameworks laid out by Aristotle and Montesquieu and constitutes a major resource from which revivalist physiopolitical theory can draw.

The Problem of Agency and Determinism

Before examining specific geopolitical arguments, it is useful to consider the interrelated issues of agency, structure, and determinism in natural-social science. How tightly are outcomes fixed by nature? What role does human agency play in shaping outcomes? And how are political institutions and structures of authority conceptualized? The answers that geopolitical theorists give to these questions are varied, and often implicit rather than explicit. But the conceptual center of gravity in the physiopolitical tradition is a functional-materialist argument distinctive from most contemporary international relations theory.

Geopolitical theorists have held a wide range of views on determinism, agency, and structure. At one extreme are those, such as the German geographer Friedreich Ratzel and the American Ellsworth Huntington, who seem to argue that there is a tight and inevitable relation between the material forces and institutional structures.[18] Humans are presented as the puppets of material forces. If a writer holding this view were perfectly consistent, it is unclear why he or she would write or act. But in practice, extremely determinist claims are often accompanied by extremely voluntaristic exhortations to more "will." Perhaps the most famous example of this combination is found in Adolf Hitler's *Mein Kampf*. At the other extreme is the French geographer Lucian Febvre's concept of "possibalism," which emphasized human freedom in responding to material forces.[19] If taken to its extreme, this view holds that material forces are of marginal and fleeting importance in explaining human affairs, more like scenic backdrops than shaping forces.

Somewhere in the middle is a view that is more difficult to conceptualize but more useful to political science. Material forces significantly define the consequences of the choices humans make, but do not dictate which choice they desire or pursue. Which institutional response(s) are in fact viable is determined, but which institutional response (if any) is made is determined by social and political forces and free agency. While it is not determined that people will make a particular response or adaptation to a particular material environment, it is inevitable that they will bear the consequences of not making an appropriate response. Human institutions are not passive before nature, the puppets of nature, or parts of nature, but it is nature and not humanity that significantly determines whether human actions achieve their goals.

This approach is illustrated by Halford Mackinder's stance toward his famous "Heartland" proposition that certain regions in the interior of Eurasia could form the basis of a world empire.[20] After analyzing material constraints and opportunities produced by the interaction between geography and industrial technology, Mackinder proposed a series of institutional remedies to prevent the emergence of a world empire. Mackinder did not argue either that a Heartland-based world empire would emerge inevitably, or that people elsewhere inevitably would take the steps necessary to avert it. Thus Mackinder's proposition can only be refuted by showing that a Heartland-based world empire does not exist *and* that institutions to prevent such an empire are not a major feature of world politics.

A related common misunderstanding about materialist and geopolitical theory derives from speaking of human institutions as "reflecting" material conditions. Such metaphors are particularly common in naturalist and materialist theories when the Marxian vocabulary of "bases" and "superstructures" is employed to conceptualize the relationship between material contexts and political structures.[21] Such mirror metaphors fail to capture the fact that naturally shaped political structures can be either *compensatory* or *exploitative*. By this I mean that political structures can either be compensations or solutions to some problem or impediment imposed by nature, or they may harness or exploit natural possibilities. Natural forces thus can shape human political institutions in diverse ways. Particular political practices and structures may be solutions or compensations for some naturally given constraint, or they may employ some naturally given asset. But whether by empowering or impeding, natural contexts shape social structures as human agents interact with them.

To sum up, in functional-materialist theory, political institutions are understood to be congealed functional practices—solution sets to meeting recurring needs in particular material contexts. Political institutions do not automatically emerge and they are not parts of nature. Rather, they are constructed by purposeful and at least partially self-conscious human agents. Although created and maintained by human agents and often built according to human designs, functional political institutions are, in their broadest features, not arbitrary or contingent in their design, but roughly succeed or fail according to their ability to achieve or perform the deeply rooted goals and tasks that motivate their construction. Thus, functional-materialism holds that much of human political activity is *practical*: purposeful, but contextualized.

Early Geopolitical Theory

Having sketched several of the main theoretical assumptions of geopolitical theory, we are now prepared to examine specific theories about the natural causes of political outcomes, specifically arguments about climate, fertile land, and the "survival of the fittest."

Climate and the Character of Nations

Prior to the industrial revolution, climate was the single most important topic in geopolitical theory.[22] The first extant work of Greek science, Hippocrates' *Airs, Waters, and Places*,[23] advanced such claims, and climate appears as a prominent variable in Aristotle, the most empirical and scientific of ancient theorists of politics, and in Montesquieu, who occupies a similarly central role in Enlightenment political science. Early climate theories typically divide the earth into temperature zones or belts and then assert that human institutions are shaped powerfully by their climatic position via the intervening variable of individual psychology. In his *Politics* Aristotle delineates three zones and argues that people in cold northern regions were prone to an excess of "spiritedness" (*thumos*) and as a result were ungovernable, but free.[24] The dissociative tendencies stemming from their spiritedness also caused the neglect of the arts and crafts (*techne*), so that northern peoples were relatively uncivilized and technologically primitive. At the other extreme, people in the torrid zones of Asia and Africa (i.e., Egypt and Lybia) were prone to an excess of bodily desire (*eros*). Pursuing the wants of the body, Asians and Africans lived in a condition of material sophistication. But the absence of spiritedness made them relatively passive politically, and more willing to accept despotic rule. Thus Asians and Africans tended to live in highly civilized despotisms.

In Aristotle's view the middle temperate zone permitted a balance between the two appetitive parts of the soul, which in turn permitted reason (*nous*), the weakest part of the psyche, to govern. The Greeks, living in the temperate zone, were thus able to strike a balance and could enjoy the benefits of material civilization without despotism. This climate geopolitics conforms with Aristotle's general emphasis on the "golden mean." He also took the argument a step further by claiming that the Greeks had a natural propensity and right to rule over lesser peoples, thus beginning a long tradition of explaining and justifying the domination of one group of humans over another by reference to natural facts.[25]

Many early modern writers, among them Bodin and Botero, extended the climate arguments of the ancients. The writings of the eighteenth-century French naturalist social scientist Montesquieu stand in much the same relation to the vast outpouring of early modern theory as Aristotle stood to his Greek predecessors. Like Aristotle, Montesquieu makes a largely unoriginal restatement of views that have subsequently come to be identified primarily with him.[26]

In attempting to understand the origins and variation in the laws of nations, Montesquieu also places a heavy emphasis on the ways in which climate affects temperament, which in turn conditions the type of custom and law of society. Unlike Aristotle, who emphasized the superiority of balanced or middle climatic conditions, Montesquieu argued that colder northern conditions are most conducive to the advance of civilization. Northern climates produce a vigorous and active way of life which in turn stimulates industry and commerce. Like Aristotle, Montesquieu linked colder climates with a love of liberty, and argued that warmer climates produce torpor, which retards material advance and induces political passivity and a predisposition toward despotism. Like Aristotle, Montesquieu did not attribute all political outcomes to climate, but rather recognized that conventions and traditions shaped by lawgivers and historical developments are also important. The arguments of both ancient and early modern climate theorists were based largely upon a limited knowledge of western Eurasia and northern Africa, and were thus flawed by an inability to assess conditions elsewhere.

Arguments about climate as an important cause of cultural, social, and political life began to decline as natural history gave way to modern social-social science in the nineteenth century. But a small number of researchers continued to advance these arguments, most notably S. F. Markham and Ellsworth Huntington.[27] Climate also plays an important part in Arnold Toynbee's elaborate theory of civilization and history. These theories of climate determinism have sought to connect variations in climate with variations in human behavior via the intervening variable of "energy," the capacity for work and labor. Markham and Huntington argue that tropical regions of the world are economically and politically retarded relative to the temperate regions because of the enervating effects of humidity and heat. To anyone from a temperate climate who has spent time in the tropics, this theory has intuitive appeal, and has been supported by studies of physiology and behavior.[28] But as a comprehensive explanation for relative levels of civilization, Huntington's theories have been broadly attacked and in the field of geography are currently seen as discredited.[29]

A major theme of twentieth-century theoreticians of climate geopolitics is that the development of manmade techniques of microclimatic control such as space heating and air conditioning expand the areas of Earth's surface habitable by "energetic" peoples. S. F. Markham argues that the expansion of civilization into the northern parts of the temperate zone resulted from the development of coal as a source of energy.[30] In this scenario, the harnessing of artificial sources of energy translates into personal and hence social vigor via the intervening variable of temperature-control technology. In this vein a largely antecdotal literature claims that the growth of the "Sunbelt" in the United States since World War II has been made possible by the new technology of microclimatic control, air conditioning.[31]

Over the last twenty-five years the study of climate's role in human affairs has become much more methodologically sophisticated and much less prone to sweeping historical generalizations. Much more accurate and detailed historical data has been assembled by demographers, archaeologists, agricultural historians, and climatologists.[32] A central interest of recent historical analysis has been with the impacts of short-term fluctuations in climate and weather. For example, the "Little Ice Age" that affected Europe between about 1500 and the middle of the nineteenth century caused widespread privation in rural areas as crops failed, and stimulated both social discontent and migration.[33]

Fertile Land and Military Insecurity

After climate, preindustrial geopolitical theorists gave greatest attention to "fertility"—arable land—as a factor in security politics. Between the close of the neolithic era and the maturation of the industrial revolution, the overwhelming majority of humankind were farmers and the main source of wealth was agricultural. Wars were primarily fought for control of arable land. Surprisingly, almost all early geopolitical theorists argued that peoples in more fertile regions suffered from grievous security problems that were directly linked to their agricultural way of life: more fertile regions are wealthier, which saps martial virtue, thus undermining military performance. In contrast, infertile regions produce poor inhabitants who make hardy warriors with incentives to plunder and who conquer richer and militarily softer peoples in fertile regions. This argument appears in Hippocrates' *Airs, Waters and Places*, Thucydides' *History of the Peloponnesian War*, Machiavelli's *Discourses*, and Gibbon's *Decline and Fall of the Roman Empire*.[34] But its most developed version appears in the writings of Ibn Khaldun, the fourteenth-century North African Islamic philosopher, whose *Muqaddimah* (Arabic for "Introduction") is widely recognized as the first comprehensive theory of history.[35] Ibn Khaldun argues that when "people settle in fertile plains and amass luxuries and become accustomed to a life of abundance and refinement, their bravery deceases."[36] In contrast, "desert life is the source of bravery, [and] savage groups are braver than others," and are therefore, "better able to achieve superiority and to take the things that are in the hands of other nations." He sees history as a cycle in which steppe peoples conquer sedentary peoples, only to themselves succumb to softening and eventual conquest (see figure 2.1).

This argument captures an important feature of premodern security politics. For most of human history, military capability and the martial virtues were closely related, because almost all combat was

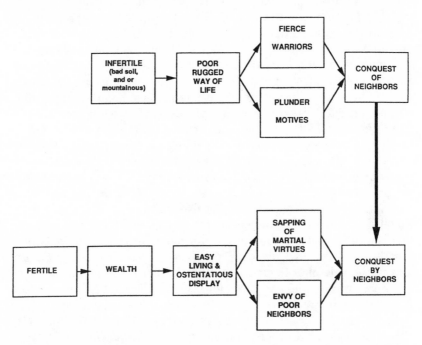

Figure 2.1 Early Views of the Military Consequences of Fertility and Infertility

at close quarters, where individual strength, endurance, bravery, and small-unit cohesion were critical to the outcome of battle. Early armies were also dependent on the feet of the soldiers for mobility, or upon horses and camels, thus placing a premium on physical endurance and conditioning. Martial virtue also was pivotal because technological changes, while important and far-reaching when they occurred, were relatively rare and were rarely monopolized by one group for long. Living close to nature, poor nomadic peoples were accustomed to deprivation and hardship, and so the rigors of war were not so onerous to them. Poor peoples in infertile regions also tended to hunt frequently and to settle their disputes violently, while sedentary peoples did not use arms or have recourse to violence in the course of their ordinary living. Prosperity served as a magnet for plunder, drawing the poor to attack the inhabitants of rich agricultural regions.

This pattern was the dominant one for many centuries, but was moderated by the military disadvantages of poverty and the countervailing advantages of wealth (see figure 2.2). Poor infertile regions supported limited populations, a product of the very infertility that kept them in poverty. But this disadvantage was not as great as the differences in aggregate population figures would suggest. Most of the population in sedentary agricultural regions was unarmed and unprepared for war because the martial arts were monopolized by elites. In contrast, virtually the entire male population of nomadic peoples were capable warriors. The smallness of population in infertile regions did, however, make it much easier for conquerors eventually to be absorbed by the sedentary peoples. The wealth of fertile areas may have been a magnet attracting plunderers and a cause of decay of military prowess, but it also had some compensating advantages: bribes could be offered,[37] mercenaries hired, and better weapons procured.

Wealthy and sedentary peoples also gradually developed institutional and cultural mechanisms to compensate for their environmental disadvantages, most notably sumptuary laws and martial education. Sumptuary laws discouraged excess private consumption and ostentatious display. In classical antiquity, both Sparta and Republican Rome were widely admired for their ability to inculcate martial virtue through civic institutions in which male citizens of all ages participated. In his famous commentary on the Roman Republic, Machiavelli observed that only "laws imposing that need to work which the situation does not impose" can prevent "idleness caused by the amenities of the land." Such laws could make "better soldiers than

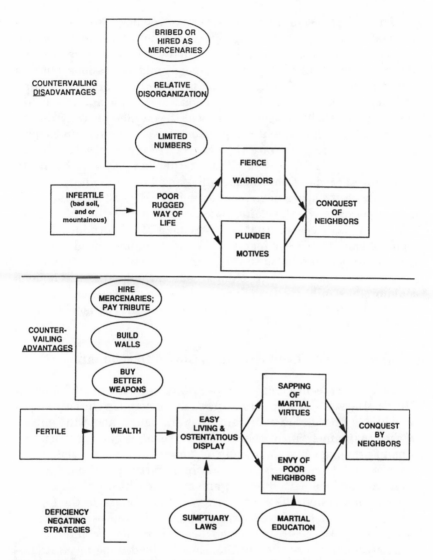

Figure 2.2 Factors and Strategies to Mitigate the Military Consequences of Fertility and Infertility

those in countries which were rough and sterile by nature," making it possible for peoples in "very fertile places" to defend themselves against attack.[38] However effective such institutions might be, they required great effort to create and maintain, and when they decayed the cycle of vulnerability and conquest resumed.

The ultimate solution to the security problem of sedentary peoples was technological. Walls and fortresses helped because the complexity of siege warfare was often beyond the abilities of nomadic peoples to master. Wealth could buy weapons and fortifications, but this was of limited value during the long periods when military technology was stagnant and the best weapons were simple and widely diffused. But the really epochal change was the development of gunpowder weapons. In his famous comparison between the Roman Empire and modern Europe, Edward Gibbon declared that gunpowder weaponry and related advances in military technology had brought an end to the ancient cycle of nomadic domination of sedentary peoples.[39]

The core of the argument connecting fertility and infertility with military outcomes is the claim that military prowess is a function of the way of life of a people that is environmentally determined. For explaining long stretches of history, this argument has considerable appeal. But this venerable naturalist argument has important limiting assumptions that are unarticulated, but variable. Infertile regions could produce military ascendancy only when close combat, employing simple pregunpowder weapons, dominated the battlefield.

Cooperation and Competition in Social Darwinism

In the late nineteenth and early twentieth centuries, the influence of Charles Darwin's theories of natural selection and evolution led to a revival and reformulation of naturalist arguments about the impact of the environment upon politics.[40] Such theories often characterize human groupings metaphorically as "organisms" whose viability is determined by their interaction with their material environment and their interactions with other organisms. But which unit of human life—the individual, the social class, the "race," state, or the human species—is analogous to Darwin's organism?[41] And which natural process—intraspecies competition or cooperation, or interspecies competition or cooperation (symbiosis)—is held to be natural? Depending on how one answers these questions, virtually any social and political practice can be vindicated as natural (see table 2.1).

By far the most influential answer to the unit-of-analysis question was the human individual. Herbert Spencer, in his influential writings on ethics and politics, began with the human individual and then proceeded to build elaborate justifications for laissez-faire capitalist society.[42] In Spencer's philosophy, so influential it has inappropriately been dubbed "social Darwinism," individual human beings

Table 2.1   Darwinian and organic analogies for human politics

UNIT OF ANALYSIS	INTERACTION	
	Competition	Cooperation
Individual Human Being	Herbert Spencer: Laissez-faire capitalism	Peter Kroptkin: "Mutual Aid" communal anarchism
Intrasocietal	Class-conflict theories	Leonard Hobhouse and Lester Ward: Social welfare state
International and Interstate	Ratzel, Kjellan: "organic states" Benjamin Kidd: "national races" Karl Pearson and others: "Aryans" biological "race" war	H. G. Wells: the human species community

are seen as competing fiercely against each other. But exactly the opposite lesson was drawn from the new evolutionary science by the Russian communal anarchist Peter Kropotkin in *Mutual Aid*: human cooperation was the key to human evolutionary success.[43] These cooperative interpretations of human social life were then further developed and applied by a variety of critics of Spencerian laissez faire, notably Leonard Hobhouse and Lester Ward.[44]

The fundamental clash between Spencer and Kropotkin reappears in many guises. In an international version of competitive Darwinism developed by Friedrich Ratzel and Rudolf Kjellen, states are portrayed as organisms in competition for scarce resources, particularly land.[45] Stripped of its vitalistic metaphysic, this line of argument resembles the varieties of realism that emphasize the competitive and zero-sum nature of interstate relations. But even if one grants that states are organisms, the behavior patterns of organisms generally are so varied that it is difficult to imagine an important state function that lacks a natural analog, in which case the organic insight loses all discriminatory and explanatory value. Another approach was to identify separate human biological "races" as the main actors. Sometimes, as in the writings of Benjamin Kidd, the term *race* is a synonym for the *nation*, rather than actual biological groupings. But in the writings of Karl Pearson and others, human biological races are understood as being in a primal contest for existence. In the hands of Adolf Hitler and the Nazi regime in Germany these racially centered Darwinian analogies served as a warrant for genocide.

In complete contrast, the prolific futurist H. G. Wells, building on both Kropotkin and Hobhouse, posited the human species itself as the basic unit of analysis and the relationship between the human species and nature as the decisive mechanism of the evolutionary struggle. Wells argued that the industrial revolution had produced a new material environment for the human species that required the rapid development of cooperative institutions on a worldwide basis in order to prevent the human species from regressing.[46]

Thus Darwinian analogies were used to justify laissez-faire capitalism, the welfare state, interstate war, racial genocide, and cosmopolitan world government. The biological sciences have a rich language and diverse set of examples, but biological analogies can never be more than interesting suggestions for understanding politics. If a compelling biological analogy constituted proof, than virtually every conceivable proposition in the human sciences could be proven. The inherent elasticity of analogic naturalism warrants a critical skepticism toward recent claims that quantum physics, cybernetics, or chaos theory vindicate any particular political practice or theory.

Nature and the Formation of the World System

Until very recently political scientists had largely abandoned the study of the effects of the natural environment on politics, but archaeologists, geographers, economists, and historians had not, and in recent decades they have made major contributions to geopolitical knowledge. Recent paleo-anthropologists have argued that natural climate change and climatic fluctuations played a key role in stimulating the evolution of humans from humanoids,[47] and the beginnings of sedentary agriculture in the Middle East.[48] Archaeologists and historians have also assembled convincing evidence that anthropogenetic ecological decay played an important role in the decline and collapse of earlier civilizations in Mesopotamia, Mesoamerica, and the Mediterranean Basin.[49]

Of more direct relevance to contemporary politics are geopolitical theories about the role of environmental factors in shaping the very uneven patterns of economic and political development during the emergence of the global system. Over the last half millennium the most far-reaching development has been the explosive emergence of

Europe as the dominant force in world history. In a burst of expansion the Europeans overran almost all the earth and created a dense system of interaction and exchange on a worldwide scale for the first time. Although direct European rule ended in the middle of the twentieth century, a "great rift" between the North and the South continues to be one of the most salient and troubling features of world politics.[50] In attempting to explain these extraordinary developments, three broad clusters of theoretical argument can be identified, two social-social scientific, and a third geopolitical. *External social systemic* explanations, including most Marxism and world-system theory, argue that the North has progressed further because it has plundered and exploited the rest of the world. Most theses on the "development of underdevelopment" emphasize that the source of the South's misery is external and located in the logic of a world system of capitalism and imperialism dominated by the North.[51] In contrast, *internalist social systemic* theories argue that domestic institutional and cultural factors explain the relative performance of these regions. Among the variants of this internalist view are the neoclassical economic claim that market institutions and incentives are the well-springs of development, and the anthropological argument that cultural dispositions toward work and savings are of decisive importance, and the political developmentalist argument that emphasizes states as catalysts of development. Despite their obvious and great differences both externalist and internalist theories look to social systemic, not natural, causes. Whether you are Karl Marx, Milton Friedman, or Gabriel Almond, variations in human institutions, not nature, are most important. A third cluster of arguments, once widely held but now pushed to a marginal status by the decay of geopolitics in political science, maintain that a great deal of the variation in performance between the West and the non-Western world can be explained as the product of ecological and climatic factors.

In order to help situate current debates about North-South environmental conflict in the context of a half millennia of environmental influences, this section will summarize historical arguments about the role of environmental factors in European development, European expansion, and the underdevelopment of tropical regions.

European Advantages

The starting point for the European ascent were changes in the patterns of economic and political activity within Europe in the early modern era that took Europe in fundamentally different directions than the other great Eurasian civilizations. Three clusters of geographical and ecological factors played an important role in these changes, and their importance is evident when contrasted with regions where they were absent (see table 2.2).

First, late Medieval and early modern Europe was geographically positioned to maximize the benefits and minimize the costs of interaction with other regions of the World Island. As a peninsula of the larger Eurasian landmass, Europe was accessible to travellers from other Eurasian civilizations (Chinese, Indian, Persian, and Arab), enabling Europe to avoid both technological and immunological stagnation. During the 1300s, for reasons not fully understood, but perhaps connected with the Mongol's unification of the interior of Eurasia, Europe was swept by the "Black Death" of bubonic plague that killed approximately a third of the population. This catastrophe contributed to the demise of feudalism and stimulated technological advance by

Table 2.2   Climatic, topographical and positional factors in early modern European history

FACTOR	CONSEQUENCES	AREA WITHOUT FACTOR
Winters interrupt pathogen transmittal	Absence of tropical diseases	Tropical Asia and Africa
Diverse climates and ecologies in relative proximity	Trade in bulk goods facilitated	Middle East, China, and India
Fragmented topography	Military consolidation impeded	Chinese river basins
High coast-to-area ratio	Trade and travel facilitated	Africa
Many rivers without waterfalls, rapids, and swamps	Trade and travel facilitated	Africa
Forests to impede horse mobility	Steppe horsemen disadvantaged	Andian and Meso-America
Human travel accessibility in Eurasia	Epidemological contact: disease endemic rather than epidemic technological diffusion	Andian and Meso-America

making labor suddenly scarce.[52] During this period, many important technologies, such as gunpowder, the compass, and mathematics came to Europe from other regions.[53] Yet at the same time, Europe was militarily buffered during the thirteenth and fourteenth centuries when all the major civilized regions of the Eurasian periphery *except* Europe were attacked by the Mongols and other nomadic peoples from the Eurasian steppes who destroyed cities, massacred the elites, and disrupted commerce. Europe was sheltered from direct military contact with the steppe peoples because the dense forests of Eastern and Central Europe could not be readily penetrated by the horsemen acclimated to the steppe environment.[54]

In contrast, the civilizations of the New World suffered from too much technological and epidemiological isolation, and were at a serious disadvantage when the Europeans suddenly penetrated their long isolation in the 1500s. The Andean and Mesomerican civilizations had isolation-induced technological lag in metallurgy, gunpowder, ships and navigation, and horses. When the European mastery of oceanic transport brought the Spaniards into contact with the Aztec and Incan empires, these disadvantages contributed to their defeat, subjugation, and cultural annihilation.

Second, the physical environment of Europe was particularly well suited to trade, and thus more likely to develop capitalism. Even Max Weber, author of the famous social-social scientific thesis on the role of the "Protestant ethic" in the formation of capitalism, observed that "the first precondition of capitalism was geographical."[55] In Europe accessibility to navigable water is unusually large, because of the numerous islands, peninsulas, and rivers that extend far inland without rapids or waterfalls. Early European capitalist activity was closely related to maritime accessibility and activity: navigable water afforded cheap transportation, thus making exchange possible on a large enough scale to justify production for the market. In the late Middle Ages, Venice, Genoa, and the Hanseatic League depended heavily upon the sea and helped pioneer capitalist practices, to be followed in the early modern era by Holland and then Britain. Europe's significant climatic diversity also contributed to opportunities for trade, because the production of many of the bulk goods for trade (e.g., wine, cork, timber, naval stores, grain, wool, and fish) was confined to specific areas by climatic conditions.[56]

In contrast, African geography and ecology were particularly unfavorable to transport and commerce, thus making economic development and capitalism much less likely there. The African continent has few islands, peninsulas, or coastal seas, and its ratio of sea coast to land area is particularly unfavorable for maritime access. Most of its major rivers have major rapids, waterfalls, and swamps that impede navigation, and several run through deep canyons and swamps infested with carnivorous reptiles, effectively isolating rather than connecting.[57]

Third, Europe's fragmented topography also had a profound effect on its political development. Prior to the industrial revolution, human mobility over land, whether for military or commercial purposes, was arduous under the best of circumstances and nearly impossible in many terrains.[58] But the landscape presented a very uneven resistance to human mobility. Mountains, forests, and bodies of water greatly impeded military mobility and thus political consolidation, such as in early Greece. Conversely, wide plains and river valleys permitted cavalry mobility, thus encouraging large-scale political consolidation, as occurred in Egypt and China. Because Europe's topography is fragmented by mountain ranges and bodies of water (such as the English Channel), separate political entities emerged in interactive proximity, but could not easily conquer one another. The particularly variegated European landscape thus tended to support a multistate system composed of smaller units than were common in the other regions of the world with comparable material civilization such as China, India, and the Middle East.[59] This meant that smaller states, where self-government was more feasible, could survive in Europe. The existence of an interactive plural state system also contributed to the viability of private property and capitalism, because the ability of capital and skilled labor to flee to more hospitable neighboring countries compelled authoritarian absolutist states to eschew high taxation and the confiscation of private property.[60] The competitive interaction of states also stimulated military technological and organizational innovation, making the European states increasingly superior to much larger imperial states elsewhere in Eurasia.[61]

Ecological Imperialism and the Disease Curtain

The internal features of the European region help explain why Europe was different, but why were the Europeans so extraordinarily successful in subjugating the rest of the world? Advantages in technology and organization were important factors, but ecological and epidemiological advantages that European conquerors and settlers enjoyed when they arrived in the New World and Oceana also played a major role.

As the historian Alfred Crosby has demonstrated, the European explorers and colonists brought with them a wide array of organisms, ranging from the black rat to the horse, that were *better adapted* to the geophysical environments of the Americas and Oceana than were species indigenous to those regions.[62] In the temperate regions of North America, southern South America, Australia and New Zealand, and far southern Africa, Eurasian pathogens brought by the Europeans had catastrophic consequences for the indigenous populations. Crosby's "ecological imperialism" has literal as well as metaphorical meaning. The most decisive ecological factor in European expansion was the impact of disease on indigenous populations. Far more AmerIndians died from diseases inadvertently introduced by the Europeans than were killed by the Europeans in battle.[63] As the indigenous population rapidly declined, the disoriented and disrupted populations were further vulnerable to direct European military attack and displacement. For the indigenous ecologies, economies, and cultures in the temperate zones outside Eurasia, the "national security" consequences of these ecological changes and exchanges were disastrous.

Not only did pathogens assault human populations, but many European plant and animal species also invaded, eliminating or marginalizing indigenous species. In the most extreme example, over the last two centuries New Zealand's ecosystem has come to be almost completely dominated by species brought, often by accident, from Northern Europe.[64] By the beginning of the twentieth century the temperate zones had become "new Europes" where European organisms and the agriculture based upon them thrived, supporting 250 million people of European descent and allowing a doubling in the percentage of all humans who were of European origin. Fueled by these ecological changes at the "great frontier," the Europe-centered world capitalist system enjoyed centuries of expansion, culminating in the great economic boom of the nineteenth century.[65] The rapid expansion and development of the United States as a "people of plenty" thus had as much to do with ecological change and displacement as with American institutions and culture.

In contrast, the trajectory of European expansion was quite different in the tropics. Here the ecological balance of power was not favorable to either European peoples or other temperate zone organisms. For several centuries, numerous diseases found only in the Tropics formed a "disease curtain" impeding European penetration of sub-Saharan Africa, the Amazon Basin, and the interiors of the islands and peninsulas of South East Asia.[66] The lag between contact and penetration was particularly extreme for Africa, which was circumnavigated by the Europeans four and a half centuries before Europeans explored and mapped its interior. It is notable that the European penetration into Africa in the middle of the nineteenth century was significantly dependent upon advances in tropical medicine.[67] But even with these advances, Europeans did not migrate to tropical regions in significant numbers. The net effect of tropical diseases and ecology was to keep the Europeans at bay for centuries, providing the technologically ill-equipped indigenous populations with the functional equivalent of a military force capable of destroying European armies equipped with advanced military technology. Because of the disease curtain Africa was subjected to direct European control last, and was subjected to European rule for less than a century.

Tropical Impediments

A related set of geopolitical arguments hold that natural impediments to economic and political development in the tropics in general, and Africa in particular, help account for the North-South divide. Advocates of this view begin by observing that the terms *North* and *South* obscure the fact that the underdeveloped South lies in the tropics, and that the few lands of temperate climate in the southern hemisphere (Argentina, Southeastern Brazil, the Union of South Africa, and Australia and New Zealand) have populations, economies, and polities similar to those found in the temperate North. In short, there is a strong correlation between temperate regions and economic development, and between tropical regions and relative underdevelopment. In attempting to explain this pattern, Andrew Kamarck, former Director of the Economic Development Institute at the World Bank, argues that high temperature and humidity hinder agriculture and handicap mineral exploration while enervating humans through disease and direct physiological restraints on strenuous labor.[68] Many diseases that debilitate hundreds of millions of people in the developing countries—malaria, schistosomiasis, trypanosomiasis, leprosy, onchocerciasis, and yellow fever—exist largely in tropical climates.

The impact of these impediments has been most severe in Africa, which is the most tropical continent and the least economically developed. A full survey of these impediments is impossible here, but one example is illustrative. Some ten million square miles of Africa (an area larger than the United States) is infested with the tsetse fly, which is a vector for diseases that kill and debilitate cattle and transport animals. These African insects are very similar to the black flies that infest northern North America, but African fly populations, unlike those in regions with killing frosts, breed continuously, thus making possible the uninterrupted transmission of infectious microorganisms.[69] The consequences for Africa's development were monumental: without animals for transport the populations of vast tracts of central Africa were isolated economically from each other and the outside world, and human slavery was highly expedient.[70]

A somewhat different version of the tropical-impediments argument has been advanced by the Tanzanian political scientist Ali Mazrui. He speaks of the "frozen ecology of capitalism," and links Africa's lag in development to climatic factors, to what he calls the "winter gap."[71] Long and harsh winters forced the Europeans to develop habits of saving and planning, but Africans, lacking such stimulus, did not need to acquire these cultural traits so central to development. In contrast to Kamarck's emphasis on environmental obstructions to wealth generation, Mazrui emphasizes the absence of the need to compensate for environmental deprivations by generating wealth in order to survive. The emphasis of one on abundance and the other on impediments to capital formation seem at odds, but can be reconciled: the tropical environment affords subsistence abundance but impedes further labor and accumulation.

The tropical-impediments argument also casts the prospects for technology transfer in a pessimistic new light. If tropical environments impose a distinct set of constraints to development, many important technologies will not "diffuse" or "transfer" from temperate to tropical environments. An implicit assumption shared by the capitalist, modernization, and unequal exchange approaches to development is that the world has a homogeneous development potential, so that only "Chinese walls of exclusion,"[72] the "cake of custom," or "unequal exchange" impede universal modernization. However, the tropical-impediments argument suggests that there are at least two populated "worlds," one temperate and one tropical. Environmental conditions are sufficiently different that some basic technologies, particularly agricultural and medicinal, will have to be largely reinvented before sustained wealth generation can occur in the tropics.[73] Not surprisingly, efforts to modernize by the introduction of systems of agriculture developed in the temperate regions into tropical regions have often been economic failures and ecological catastrophes. The peculiar ecology of the Amazon rain forest has repeatedly frustrated even the most well-funded and ambitious agricultural and silvicultural schemes.[74] These failures have led many ecologists to emphasize the importance of building upon local "agro-ecological" approaches instead of importing temperate zone forestry and agricultural systems.[75]

Conclusions

This brief analysis, sampling, and survey has mapped some of the promise and problems of bringing nature back into social science. For political scientists and security analysts, this project has just begun, and much work—both conceptual and empirical—remains to be done. This essay has illuminated some of the conceptual moves and pitfalls that attend this enterprise, and has demonstrated that geopolitical approaches have both a long history and considerable explanatory power in explaining important political events of the past. Three observations by way of conclusion emerge from this analysis.

First, to provide a robust analysis of the impact of natural factors and human-nature interaction, social science must supplement its concentration on social causes of social outcomes by returning to the natural-social scientific approach of the geopolitical tradition. Simply injecting natural variables into existing social scientific models is not likely to be very fruitful. Rather, a far-reaching reconceptualization of the role of nature in social practice is needed. Both logic and past efforts demonstrate that analogic naturalism is a perennially seductive but inevitably fruitless avenue, and that functional structural approaches promise better results. Unfortunately, the recent turn of social theory away from structuralism to an emphasis on the socially constructed nature of reality, while offering many insights and correctives to mechanistic structural arguments, could exacerbate the incapacity of social science to deal with environmental issues. Social structures are constructed by humans rather than themselves being natural. But social practices rise and fall and are valued or rejected not solely—or even primarily—because of socially constructed criteria, but rather because of their ability to function successfully in meeting enduring human needs in material contexts that are both diverse and shifting. Social constructions inevitably shape how nature is perceived and acted upon, but nature itself is not socially constructed, and any social science that assumes so will inevitably be blind to important aspects of human life. Nature constitutes a structuring reality for human beings that is not socially constructed. In short, a return to functional-materialist theory is the key to bringing nature back into social theory.

The second conclusion that emerges from this chapter is that geopolitical theories of conflict and cooperation have a long and diverse lineage, and defy easy classification. Earlier geopolitical theories of the impact of arable land upon security exhibit both the strengths and weaknesses of geopolitical theorizing. Here the conventional wisdom among analysts of environmental security is challenged, forcing the realization that the ways in which natural factors impinge upon the performance of security institutions is not straightforward and is often counterintuitive.

Third and finally, the synthetic reconstruction of a geopolitical explanation for the rise of Europe and the great North-South divide demonstrate that the legacies of environmental factors on politics are both profound and long lasting. Here geopolitical theory should serve to deflate the self-confidence and arrogance of the currently hegemonic Western civilization by showing that its rise and influence owes as much to contingent and accidental natural factors as to institutional innovation. At the same time, this geopolitical account undermines the optimistic and progressive assumptions of the Third World left that uneven and failed development is going to be solved by an institutional agenda of redistribution and equal empowerment.

Looking to the future with this naturalist understanding of the past and present, we are left wondering where the next turn of nature's wheel of fortune will leave the world. On the basis of what has come before, we can be confident only that the ways in which mother nature shapes the fates of her human children will be as powerful, as complicated, and as unpredictable as they have always been.

–––––––––––––––––––––––––– NOTES ––––––––––––––––––––––––––

Earlier versions of this chapter were presented at the conference on "Environmental Change and Security," Institute of International Relations, The University of British Columbia, Vancouver. April 29 to May 1, 1993, and the American Political Science Association meeting, September 1993, Washington, D.C. The author gratefully acknowledges extensive helpful comments from John Agnew, Ken Conca, and Peter Haas.

1. Harold and Margaret Sprout, *Towards a Politics of the Planet Earth* (New York: Van Nostrand Reinhold, 1971); Dennis Pirages, *Global Ecopolitics: The New Context for International Relations* (North Scituate, Mass.: Duxbury Press, 1978); and Dennis Pirages, *Global Technopolitics: The Inter-*

national Politics of Technology and Resources (Pacific Grove, Calif.: Brooks/ Cole, 1989).

2. Clarence Glacken, *Traces on the Rhodian Shore: Nature and Culture in Western Thought from Ancient Times to the End of the Eighteenth Century* (Berkeley: University of California Press, 1967).

3. Montesquieu, *Spirit of the Laws* [original 1748] (Cambridge: Cambridge University Press, 1989), book 19, section 14.

4. Stephen M. Walt, "The Renaissance of Security Studies," *International Studies Quarterly* vol. 35, no. 2 (1991), pp. 211–39.

5. For overviews, see Ladis Kristof, "The Origins and Evolution of Geopolitics," *Journal of Conflict Resolution* vol. 4 (March 1960); and David Livingstone, *The Geographical Tradition* (Oxford: Basil Blackwell, 1992).

6. Not all theories about the relationship between humanity and nature are naturalist (ie., subscribe cause to nature). My schema thus does not encompass claims about the impact of humans upon nature, nor claims about the ethical value or "rights" of nature.

7. James Lovelock, *Gaia: A New Look at Life on Earth* (New York: Oxford University Press, 1979).

8. Charles Darwin, *The Origin of Species* [original 1859] (New York: Modern Library, n.d.), p. 117.

9. Melvin Konner, *The Tangled Wing: Biological Constraints on the Human Spirit* (New York: Harper & Row, 1982); Mary Midgley, *Beast and Man: The Roots of Human Nature* (Ithaca, N.Y.: Cornell University Press, 1978).

10. Roger Masters, *The Nature of Politics* (New Haven, Conn.: Yale University Press, 1989); and Mary Maxwell, *Morality among Nations: An Evolutionary View* (Albany: State University of New York Press, 1990).

11. Marvin Harris, *Cultural Materialism: The Struggle for a Science of Culture* (New York: Random House, 1979); and Martin F. Murphy and Maxine L. Margolis, eds., *Science, Materialism, and the Study of Culture* (Gainesville: University Press of Florida, 1995).

12. For analysis of the problems of technology and change in functional-material theory, see Daniel Deudney, "Geopolitics and Change," in Michael Doyle and G. John Ikenberry, eds., *New Thinking in International Relations Theory* (Boulder, Colo.: Westview, 1997), pp. 91–124.

13. Samuel Z. Klausner, "Social Science and Environmental Quality," *The Annals, of the American Academy of Political and Social Science* vol. 389 (May 1970), p. 4.

14. Kenneth Waltz, *Man, the State and War* (New York: Columbia University Press, 1959).

15. Frederick Teggart, *Processes of History* (New Haven, Conn.: Yale University Press 1918), pp. 19, 23.

16. Geoffrey Barraclough, *Main Trends in History* (expanded and updated by Michael Burns) (London: Holmes & Meier, 1991), p. 158.

17. Fernand Braudel, *On History*, trans. Sarah Matthews (Chicago: University of Chicago Press, 1980), p. 185.

18. Friedreich Ratzel, *Politische Geographie* (Munich: Oldenburg, 1879).

19. Lucian Febvre, *A Geographical Introduction to History* (London: Kegan, Paul, 1925); O. H. K. Spate, "How Determined Is Possibilism?" *Geographical Studies* vol. 4 (1957); Harold and Margaret Sprout, "Environmental Determinism," "Free-will Determinism," and "Possibilism," in *The Ecological Perspective on Human Affairs* (Princeton, N.J.: Princeton University Press, 1965); George Tatham, "Environmentalism and Possibilism," in Griffith Taylor, ed., *Geography in the Twentieth Century* (New York: Philosophical Library, 1951).

20. Halford Mackinder, *Democratic Ideals and Reality* (London: Holt, 1919).

21. Raymond Williams, "Determination" and "Base and Superstructure," in *Marxism and Literature* (New York: Oxford University Press, 1977), pp. 83–89.

22. Glacken, *Traces on the Rhodian Shore*; Capt. Charles Konigsberg, "Climate and Society: A Review of the Literature," *Journal of Conflict Resolution* vol. 4 (March 1960).

23. Hippocrates, "Airs, Waters and Places," in G. E. R. Lloyd, ed., *Hippocratic Writings* (London: Penguin, 1978).

24. Aristotle, *The Politics*, Barker ed. (Oxford: Oxford University Press, 1948), paragraph 1327, p. 298.

25. For further discussion, see: David Livingstone, *The Geographical Tradition* (Oxford: Blackwell, 1993), pp. 1–32, 216–59; and David Arnold, *The Problem of Nature: Environment, Culture and European Expansion* (Oxford: Blackwell, 1996).

26. Montesquieu, *Spirit of the Laws*, books 14–18.

27. Ellsworth Huntington, *Civilization and Climate* (New Haven, Conn.: Yale University Press, 1915); Ellsworth Huntington, *The Mainsprings of Civilization* (New York: John Wiley and Sons, 1945).

28. Clarence A. Mills, "Temperature Dominance over Human Life," *Science* vol. 110 (September 16, 1949).

29. Geoffrey Martin, "The Geography of Ellsworth Huntington: Some Thoughts and Reflections," ch. 14 of *Ellsworth Huntington: His Life and Thought* (Hamden, Conn.: Archon Books, 1973); and Stephen Visher, "Ellsworth Huntington: Human Ecologist," *Journal of Human Ecology* vol. 2 (1952), pp. 1–27.

30. F. S. Markham, *Climate and the Energy of Nations* (London: Oxford University Press, 1947).

31. Robert Friedman, "The Air-Conditioned Century," *American Heritage* (August/September 1984), pp. 20–32.

32. For example, see: T. M. L. Wigley and G. Farmer, eds., *Climate and History: Studies in Past Climates and Their Impact on Man* (Cambridge, U.K.: Cambridge University Press, 1981); and Robert I. Rotberg and Theodore K. Rabb, eds., *Climate and History: Studies in Interdisciplinary History* (Princeton, N.J.: Princeton University Press, 1981).

33. H. H. Lamb, *Climate, History and the Modern World* (London: Methuen, 1982), pp. 162–252; and J. M. Grove, *The Little Ice Age* (London: Methuen, 1988).

34. Thucydides, *The Peloponnesian War*, trans. Rex Warner (Baltimore: Pelican, 1974); and Edward Gibbon, ch. 26, *The Decline and Fall of the Roman Empire* (New York: Modern Library, n.d.) [original 1776].

35. Ibn Khaldun, *The Muqaddimah: An Introduction to History* [original 1377] (Princeton, N.J.: Princeton University Press, 1967).

36. Ibn Khaldun, *Muqaddimah*, p. 107.

37. Bribes played a particularly important role in the Chinese strategy to regulate the predations of the steppe peoples. Lec Kwanten, *Imperial Nomads: A History of Central Asia, 500–1500* (Philadelphia: University of Pennsylvania Press, 1979).

38. Niccolo Machiavelli, *The Discourses*, trans. Walker and Richardson (Baltimore, Md.: Penguin, 1970), pp. 102–103.

39. Edward Gibbon, *The Decline and Fall*, pp. 436–44.

40. For overviews, see Robert Mackintosh, *From Comte to Benjamin Kidd: The Appeal of Biology or Evolution for Human Guidance* (New York: 1899); and Paul Crook, *Darwinism, War and History: The Debate over the Biology of War from the Origins of Species to the First World War* (Cambridge, U.K.: Cambridge University Press, 1994).

41. Johannes Mattern, *Geopolitik: Doctrine of National Self-Sufficiency and Empire* (Baltimore, Md.: Johns Hopkins University Press, 1942).

42. Herbert Spencer, *Man Versus the State* (London: Watts, 1892).

43. Peter Kropotkin, *Mutual Aid* (London: Macmillan, 1903).

44. Leonard Hobhouse, *Social Evolution and Political Theory* (New York: Columbia University Press, 1911); and Lester Ward, "The Laissez Faire Doctrine Is Suicidal" [1884] and "Social and Biological Struggles" [1906], in Henry Steele Commager, ed, *Lester Ward and the Welfare State* (Indianapolis, Ind.: Bobbs-Merrill, 1967), pp. 63–68, 391–402.

45. Robert Strauss-Hupe, *Geopolitics: The Struggle for Space and Power* (New York: G. P. Putnam's Sons, 1942); and Andreas Dorpalan, ed., *The World of General Haushofer: Geopolitics in Action* (New York: Farrar and Rinehart, 1942).

46. H. G. Wells, "Human Evolution, an Artificial Process," *Fortnightly Review* vol. 66 (1896), pp. 590–95; and W. Warren Wager, *H. G. Wells and the World State* (New Haven, Conn.: Yale University Press, 1961).

47. Yves Coppens, "East Side Story: The Origin of Humankind," *Scientific American* vol. 270, no. 5 (May 1994), pp. 88–95.

48. Donald O. Henry, *From Foraging to Agriculture: The Levant at the End of the Ice Age* (Philadelphia: University of Pennsylvania Press, 1989).

49. For a good summation, unfortunately not footnoted, see Clive Ponting, *A Green History of the World* (New York: St. Martin's, 1992).

50. L. S. Stavrianos, *Global Rift* (New York: Morrow, 1981).

51. For a succinct overview of neo-Marxist world-system theory, see Thomas Richard Shannon, *An Introduction to the World-System Perspective* (Boulder, Colo.: Westview Press, 1989).

52. Robert S. Gottfried, *The Black Plague: Natural and Human Disaster in Medieval Europe* (New York: Free Press, 1983).

53. Arnold Pacey, *Technology in World Civilization* (Cambridge, Mass.: MIT Press, 1990).

54. Perry Anderson, *Lineages of the Absolutist State* (London: New Left Books, 1974).

55. Max Weber, *General Economic History* (Glencoe, Ill.: Free Press, 1927), p. 353.

56. ". . . the difference of climate is the cause that several nations have great occasion for the merchandise of each other." Montesquieu, *Spirit of the Laws*, book 21, section 5; and E. L. Jones, *The European Miracle: Environments, Economics, and Geopolitics in the History of Europe and Asia* (Cambridge, U.K.: Cambridge University Press, 1981).

57. Andrew M. Kamarck, *The Economics of African Development* (New York: Praeger, 1971).

58. James E. Vance, Jr., *Capturing the Horizon: The Historical Geography of Transportation* (New York: Harper and Row, 1986).

59. Montesquieu, *Spirit of the Laws*, book 17, section 6; and E. L. Jones, *The European Miracle*.

60. Montesquieu, *Spirit of the Laws*, book 21, section 5; and Charles Tilly, *Coercion, Capital, and European States, A.D. 990–1992* (Oxford: Basil Blackwell, 1990).

61. William McNeill, *The Pursuit of Power: Technology, Armed Force and Society since A.D. 1000* (Chicago: University of Chicago Press, 1982).

62. Alfred Crosby, *Ecological Imperialism: The Biological Expansion of Europe, 900–1900* (Cambridge, U.K.: Cambridge University Press, 1986).

63. Percy Ashburn, *The Ranks of Death: A Medical History of the Conquest of America* (New York: Coward and McCann, 1947); Alfred Crosby, *The Columbian Exchange: Biological and Cultural Consequences of 1492* (Austin: University of Texas Press, 1972); William McNeill, *Plagues and Peoples* (New York: Morrow, 1976).

64. Alfred Crosby, *Ecological Imperialism*, pp. 217–268.

65. Walter Prescott Webb, *The Great Frontier* (Austin: University of Texas Press, 1951).

66. Philip Curtin, *Death by Migration* (Madison: University of Wisconsin Press, 1989); and Philip Curtin, *The Image of Africa: British Ideas and Actions, 1780–1964* (Madison: University of Wisconsin Press, 1964).

67. Donald R. Headrick, *The Tools of Empire: Technology and European Imperialism in the Nineteenth Century* (New York: Oxford University Press, 1981), pp. 58–79.

68. Andrew M. Kamarck, *The Tropics and Economic Development: A Provocative Inquiry into the Poverty of Nations* (Baltimore, Md.: Johns Hopkins University Press, 1976), p. 3; Don L. Lambert, "The Role of Climate in the Economic Development of Nations," *Land Economics* vol. 47 (1971), pp. 339–417.

69. Kamarck, *The Topics and Economic Development*, p. 59.

70. Kamarck, *The Topics and Economic Development*, p. 38.

71. Ali Mazrui, *The Africans: A Triple Heritage* (Boston: Little, Brown, 1986), pp. 216–23.

72. Sholomo Avineri, ed., *Karl Marx on Colonialism and Modernization*

(Garden City, N.Y.: Doubleday Anchor, 1969). For a critique of physicalist theories from a Marxist perspective, see J. M. Blaut, *The Colonizer's Model of the World: Geographical Diffusion and Eurocentric History* (New York: Guilford Press, 1993).

73. Kenneth Boulding, "Is Economics Culture Bound?" *American Economic Review* vol. 60 (May 1970).

74. Susanna Hecht and Alexander Cockburn, *The Fate of the Forest* (New York: HarperCollins, 1990), pp. 87–103.

75. Juan de Onis, *The Green Cathedral: Sustainable Development of the Amazon* (New York: Oxford University Press, 1992).

PART II

The Contemporary Debate

3

Thresholds of Turmoil

Environmental Scarcities and Violent Conflict

Thomas F. Homer-Dixon

Within the next fifty years, the planet's human population will probably pass nine billion, and global economic output may quintuple. Largely as a result of these two trends, scarcities of renewable resources will increase sharply. The total area of high-quality agricultural land will drop, as will the extent of forests and the number of species they sustain. Coming generations will also see the widespread depletion and degradation of aquifers, rivers, and other water resources; the decline of many fisheries; and perhaps significant climate change.

If such "environmental scarcities" become severe, could they precipitate violent civil or international conflict? In a previous article, I surveyed the issues and evidence surrounding this question and proposed an agenda for further research.[1] In this chapter I report the results of an international research project guided by this agenda.[2] I first review my original hypotheses and the project's research design. I then present several general findings of this research that led me to revise the original hypotheses. The chapter continues with an assessment of empirical evidence for and against the revised hypotheses. It concludes with an account of scarcity's various roles as a cause of conflict and an assessment of the implications of environmentally induced conflict for international security.

Our research showed that environmental scarcities are already contributing to violent conflicts in many parts of the developing world. These conflicts are probably the early signs of an upsurge of violence in the coming decades that will be induced or aggravated by scarcity. The violence will usually be subnational, persistent, and diffuse. Poor societies will be particularly affected since they are less able to buffer themselves from environmental scarcities and the social crises they cause. These societies are, in fact, already suffering acute hardship from shortages of water, forests, and especially fertile land.

Social conflict is not always a bad thing. Mass mobilization and civil strife can produce opportunities for beneficial change in the distribution of land and wealth and in processes of governance. But fast-moving, unpredictable, and complex environmental problems can overwhelm efforts at constructive social reform. Moreover, scarcity can sharply increase demands on key institutions, such as the state, while it simultaneously reduces their capacity to meet those demands. These pressures increase the chance that the state will either fragment or become more authoritarian. The negative effects of severe environmental scarcity are therefore likely to outweigh the positive.

General Findings

Our research was intended to provide a foundation for further work. We therefore focused on two key preliminary questions: Does environmental scarcity cause violent conflict? And, if it does, how does it operate?

The research was structured as I had proposed in my previous work. Seven types of environmental change were identified as plausible causes of violent intergroup conflict: greenhouse-induced climate change, stratospheric ozone depletion, acid deposition, degradation and loss of good agricultural land, degradation and removal of forests, depletion and pollution of fresh water supplies, and depletion of fisheries. We used three hypotheses to link these changes with violent conflict. First, we proposed that increasing shortages of physically controllable environmental resources—such as clean water and good agricultural land—provoke interstate "simple-scarcity" conflicts, or, in common parlance, resource wars. Second, we hypothesized that large population movements caused by environmental stress induce "group-identity" conflicts, especially ethnic clashes. And third, we proposed that severe environmental scarcity simultaneously increases economic deprivation and disrupts key social institutions, which in turn causes "deprivation" conflicts such as civil strife and insurgency.

Two detailed case studies were completed for each of the three research hypothesis.[3] By selecting cases that appeared, prima facie, to show a link between environmental change and conflict, we sought to falsify the null hypothesis that environmental scarcity does not cause violent conflict. By carefully tracing the causal processes in each case, we also sought to identify how environmental scarcity operates if and when it is a cause. The completed case studies were reviewed at a series of expert workshops; in light of these findings, I revised the original hypotheses and identified common variables and processes across the cases. These conclusions were reviewed by the project's core team of experts.

The following are four general findings of this research.

1. Of the major environmental changes facing humankind, degradation and depletion of agricultural land, forests, water, and fish will contribute more to social turmoil in coming decades than climate change or ozone depletion.

When analysts and policymakers in developed countries consider the social impacts of large-scale environmental change, they focus undue attention on climate change and stratospheric ozone depletion. But vast populations in the developing world are already suffering from shortages of good land, water, forests, and fish; in contrast, the social effects of climate change and ozone depletion will probably not be seen until well into the next century. If these atmospheric problems do eventually have an impact, they will most likely operate, not as individual environmental stresses, but in interaction with other, long-present resource, demographic, and economic pressures that have gradually eroded the buffering capacity of some societies.

Mexico, for example, is vulnerable to such interactions. People are already leaving the state of Oaxaca because of drought and soil erosion. Researchers estimate that future global warming could decrease Mexican rainfed maize production up to 40 percent. This change could in turn interact with ongoing land degradation, free trade (Mexico's comparative advantage is in water-intensive fruits and vegetables), and the privatization of communal peasant lands to cause grave internal conflict.[4]

2. Environmental change is only one of three main sources of scarcity of renewable resources; the others are population growth and unequal social distribution of resources. The concept "environmental scarcity" encompasses all three sources.

Analysts often usefully characterize environmental problems as resource scarcities. Resources can be roughly divided into two groups: nonrenewables, like oil and iron ore, and renewables, like fresh water, forests, fertile soils, and the earth's ozone layer. The latter category includes renewable "goods" such as fisheries and timber, and renewable "services" such as regional hydrological cycles and a benign climate.

The commonly used term *environmental change* refers to a human-induced decline in the quantity or quality of a renewable resource that occurs faster than it is renewed by natural processes. But this concept limits the scope of environment-conflict research. Environmental change is only one of three main sources of renewable resource scarcity. The second, population growth, reduces a resource's per capita availability by dividing it among more and more people. The third, unequal resource distribution, concentrates a resource in the hands of a few people and subjects the rest to greater scarcity.[5] The property rights that govern resource distribution often change as a result of large-scale development projects or new technologies that alter the relative values of resources.

A simple "pie" metaphor illustrates the three sources of scarcity: reduction in the quantity or quality of a resource shrinks the resource pie, while population growth divides the pie into smaller slices for each individual, and unequal resource distribution means that some groups get disproportionately large slices.[6] Unfortunately, analysts often study resource depletion and population growth in isolation from the political economy of resource distribution. The term *environmental scarcity*, however, allows these three distinct sources of scarcity to be incorporated into one analysis. Empirical evidence suggests, in fact, that the first two sources are most pernicious when they interact with unequal resource distribution.

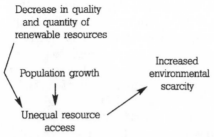

Figure 3.1(a)   Resource Capture: Resource Depletion and Population Growth Cause Unequal Resource Access

3. The three sources of environmental scarcity often interact, and two patterns of interaction are particularly common: "resource capture" and "ecological marginalization."

There are two principal patterns of interaction among the three sources of scarcity (see figure 3.1). First, a fall in the quality and quantity of renewable resources can combine with population growth to encourage powerful groups within a society to shift resource distribution in their favor. This can produce dire environmental scarcity for poorer and weaker groups whose claims to resources are opposed by these powerful elites. I call this type of interaction "resource capture." Second, unequal resource access can combine with population growth to cause migrations to regions that are ecologically fragile, such as steep upland slopes, areas at risk of desertification, and tropical rain forests. High population densities in these areas, combined with a lack of knowledge and capital to protect local resources, causes severe environmental damage and chronic poverty. This process is often called "ecological marginalization."[7]

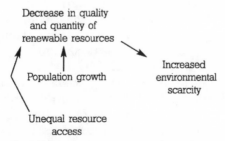

Figure 3.1(b)   Ecological Marginalization: Unequal Resource Access and Population Growth Cause Resource Degradation and Depletion

Events in the Senegal River Valley in 1989 illustrate resource capture. The valley demarcates the border between Senegal and Mauritania in West Africa. Senegal has fairly abundant agricultural land, but much of it suffers from high to severe wind and water erosion, loss of nutrients, salinization because of overirrigation, and soil compaction caused by intensification of agriculture.[8] The country has an overall population density of thirty-eight people per square kilometer and a population growth rate of 2.8 percent; in twenty-five years the population will double.[9] In contrast, except for the Senegal Valley along its southern border and a few oases, Mauritania is largely arid desert and semiarid grassland. Its population density is very low at about two people per square kilometer, but the growth rate is 2.9 percent. This combination of factors led the Food and Agriculture Organization (FAO) and two other organizations in a 1982 study to include both Mauritania and Senegal in their list of "critical" countries whose croplands cannot support their current and projected populations without a large increase in agricultural inputs, such as fertilizer and irrigation.[10]

Normally, the broad floodplains fringing the Senegal River support productive farming, herding, and fishing based on the river's annual floods. During the 1970s, however, the prospect of chronic food shortages and a serious drought encouraged the region's governments to seek international financing for the high Manantali Dam on the Bafing River tributary in Mali and the Diama salt-intrusion barrage near the mouth of the Senegal River between Senegal and Mauritania. These dams were designed to regulate the river's flow to produce hydropower, expand irrigated agriculture, and provide riverine transport from the Atlantic Ocean to landlocked Mali, which lies to the east of Senegal and Mauritania.

But the plan had unfortunate and unforeseen consequences. Anticipation of the new dams sharply increased land values along the river in areas where high-intensity agriculture was to become feasible. The elite in Mauritania, which consists mainly of white Moors, then rewrote legislation governing land ownership, effectively abrogating the rights of black Africans to continue farming, herding, and fishing along the Mauritanian riverbank.[11]

There has been a long history of racism by white Moors in Mauritania toward their non-Arab, black compatriots. In the spring of 1989, the killing of Senegalese farmers by Mauritanians in the river basin triggered explosions of ethnic violence in the two countries. In Senegal, almost all of the 17,000 shops owned by Moors were destroyed, and their owners were deported to Mauritania. In both countries several hundred people were killed and the two nations nearly came to war. The Mauritanian regime used this occasion to activate the new land legislation, declaring the Mauritanians who lived alongside the river to be "Senegalese," thereby stripping them of their citizenship; their property was seized. Some 70,000 of the black Mauritanians were forcibly expelled to Senegal, from where some launched raids to retrieve expropriated cattle. Diplomatic relations between the two countries have now been restored, but neither has agreed to allow the expelled population to return or to compensate them for their losses.

We see here the interaction of two sources of human-induced environmental scarcity. Degradation of the land resource and population pressures helped precipitate agricultural shortfalls, which in turn encouraged a large development scheme. These factors together raised land values in one of the few areas in either country offering the potential for a rapid move to high-intensity agriculture. A powerful elite then changed property rights and resource distribution in its favor, which produced a sudden increase in resource scarcity for an ethnic minority, expulsion of the minority, and ethnic violence.

The Philippines offers a good illustration of ecological marginalization. Unequal access to rich agricultural lowlands combines with population growth to cause migration to easily degraded upland areas; erosion and deforestation contribute to economic hardship that spurs insurgency and rebellion.

Spanish and American colonial policies in the Philippines left behind a grossly unfair distribution of good cropland in lowland regions—an imbalance perpetuated since independence by a powerful landowning elite. Since World War II, green revolution technologies have greatly increased lowland production of grain for domestic consumption and of cash crops—like sugar, coconut, pineapple, and bananas—that help pay the country's massive external debt. This has raised demand for agricultural labor on large farms, but not enough to compensate for a population growth rate of 2.5 to 3.0 percent. Together, therefore, unequal land access and population growth have produced a surge in agricultural unemployment.

With insufficient rural or urban industrialization to employ this excess labor, there has been unrelenting downward pressure on wages. Economically desperate, millions of poor agricultural laborers and landless peasants have migrated to shantytowns in already overburdened cities, such as Manila; millions of others have moved to the least productive—and often most ecologically vulnerable—territories, such as steep hillsides.[12] In these uplands, settlers use fire to clear forested or previously logged land. They bring with them little money or knowledge to protect their fragile ecosystems, and their small-scale logging, production of charcoal for the cities, and slash-and-burn farming often cause horrendous environmental damage—particularly water erosion, landslides, and changes in the hydrological cycle.[13] This has set in motion a cycle of falling food production, clearing of new plots, and further land degradation. There are few new areas in the country that can be opened up for agricultural production, so even marginally fertile land is becoming hard to find in many places, and economic conditions are often critical for the peasants.[14]

The situation in the Philippines is not unique. Ecological marginalization occurs with striking regularity around the planet, affecting hundreds of millions of people in places as diverse as the Himalayas, Indonesia, Costa Rica, Brazil, and the Sahel.

4. Societies can use many strategies to adapt to environmental scarcity; these strategies will involve either continued reliance on indigenous resources or "decoupling" from these resources. In both cases, the supply of social and technical ingenuity must be adequate.

Societies will better avoid turmoil if they can adapt to environmental scarcity so that it does not cause great suffering. Strategies for adaptation fall into two categories. First, societies can continue to rely on their indigenous resources but use them more sensibly and provide alternative employment to people who have limited resource access. For example, economic incentives like increases in resource prices and taxes can reduce degradation and depletion by encouraging conservation, technological innovation, and resource substitution. Family planning and literacy campaigns can ease population-growth-induced scarcity. Land redistribution and labor-intensive rural industries can relieve the effects of unequal access to good cropland.

Second, the country might be able to "decouple" itself from dependence on its own depleted environmental resources by producing goods and services that do not rely heavily on those resources; the country could then trade the products on the international market for the resources it no longer has at home. The decoupling might, in fact, be achieved by rapidly exploiting the country's environmental resources and reinvesting the profits in capital, industrial equipment, and skills to permit a shift to other forms of wealth creation. For instance, Malaysia can use the income from overlogging its forests to fund a modern university system that trains electrical engineers and computer specialists for a high-technology industrial sector.

If either strategy is to succeed, a society must be able to supply enough ingenuity at the right places and times. Two kinds are key. Technical ingenuity is needed to develop, for example, new agricultural and forestry technologies that compensate for environmental loss. Social ingenuity is needed to create institutions and organizations that buffer people from the effects of scarcity and provide the right incentives for technological entrepreneurs. Social ingenuity is therefore often a precursor to technical ingenuity. The development and distribution of new grains adapted for dry climates and eroded soils, of alternative cooking technologies to compensate for the loss of firewood, and of water conservation technologies, depends on an intricate and stable system of markets, legal regimes, financial agencies, and educational and research institutions.

In the next decades, the need for both technical and social ingenuity to deal with environmental scarcities will rise sharply. Population growth, rising average resource consumption, and persistent inequalities in resource access ensure that scarcities will affect many environmentally sensitive regions with a severity, speed, and scale unprecedented in history. Resource-substitution and conservation tasks will be more urgent, complex, and unpredictable, driving up the need for technical ingenuity. Moreover, solving these problems through market and other institutional innovations (such as changes in property rights and resource distribution) will require great social ingenuity.

At the same time that environmental scarcity is boosting the demand for ingenuity, it may interfere with supply. Poor countries start at a disadvantage: they are underendowed with the social institutions—including the productive research centers, efficient markets, and capable states —that are necessary for an ample supply of both social and technical solutions to scarcity. Moreover, their ability to create and maintain these institutions may be diminished by the very environmental stress they need to address, because scarcity can weaken states, as we shall see, and it can engender intense rivalries between interest groups and elite factions.

Evidence Bearing on the Hypotheses

The above findings led me to revise the original three hypotheses by redefining the independent variable, "environmental scarcity." I narrowed the range of environmental problems hypothesized to cause conflict, deemphasizing atmospheric problems, and focusing instead on forests, water, fisheries, and especially cropland. I expanded the scope of the independent variable to include scarcity caused by population growth and resource maldistribution as well as that caused by degradation and depletion. And I also incorporated into the variable the role of interactions between these three sources of scarcity.

Our research project produced the following empirical evidence bearing on the three hypotheses thus revised.

Hypothesis One: Environmental scarcity causes simple-scarcity conflicts between states.

There is little empirical support for this first hypothesis. Scarcities of renewable resources do not often cause resource wars between states. This finding is intriguing because resource wars have been common since the beginning of the state system. For instance, during World War II, Japan sought to secure oil, minerals, and other resources in China and Southeast Asia; and the Gulf War was at least partly motivated by the desire for oil.

However, we must distinguish between nonrenewable and renewable resources. Arthur Westing has compiled a list of twelve conflicts in the twentieth century involving resources, beginning with World War I and concluding with the Falklands/Malvinas War.[15] Access to oil and/or minerals was at issue in ten conflicts. Five conflicts involved renewable resources, while only two—the 1969 Soccer War between El Salvador and Honduras and the Anglo-Icelandic Cod War of 1972–1973—concerned neither oil nor minerals. Cropland was a factor in the former case, and fish in the latter. But the Soccer War was not a simple-scarcity conflict between states; rather it arose from the ecological marginalization of Salvadorean peasants and their consequent migration into Honduras.[16] It is supporting evidence, therefore, for our second and third hypotheses, but not for the first. And, since the Cod War involved negligible violence, it hardly qualifies as a resource war.

States have fought more over nonrenewable than renewable resources for two reasons, I believe. First, petroleum and mineral resources can be more directly converted into state power than agricultural land, fish, and forests. Oil and coal fuel factories and armies, and ores are vital for tanks and naval ships. In contrast, although captured forests and cropland may eventually generate wealth that can be harnessed by the state for its own ends, this outcome is more remote in time and less certain. Second, the very countries that are most dependent on renewable resources, and which are therefore most motivated to seize resources from their neighbors, also tend to be poor, which lessens their capability for aggression.

Our research suggests that the renewable resource most likely to stimulate interstate resource war is river water.[17] Water is a critical resource for personal and national survival; furthermore, since river water flows from one area to another, one country's access can be affected by another's actions. Conflict is most probable when a downstream riparian is highly dependent on river water and is strong in comparison to upstream riparians. Downstream riparians often fear that their upstream neighbors will use water as a means of leverage. This situation is particularly dangerous if the downstream country also believes it has the military power to rectify the situation. The relationships between South Africa and Lesotho and between Egypt and Ethiopia have this character.

The Lesotho case is interesting. Facing critical water shortages, South Africa negotiated in vain with Lesotho for thirty years to divert water from the Kingdom's mountains to arid Transvaal. In 1986 South Africa gave decisive support to a successful military coup against Lesotho's tribal government. South Africa stated that it helped the coup because Lesotho had been providing sanctuary to guerrillas of the African National Congress. This was undoubtedly a key motivation, but within months the two governments reached agreement to construct the huge Highlands Water Project to meet South Africa's needs. It seems likely, therefore, that the desire for water was an ulterior motive behind South African support for the coup.[18]

But our review of the historical and contemporary evidence shows that conflict and turmoil related to river water are more often internal than international. The huge dams that are often built to deal with general water scarcity are especially disruptive. Relocating large numbers of upstream people generates turmoil among the relocatees and clashes with local groups in areas where the relocatees are resettled. The people affected are often members of ethnic or minority groups outside the power hierarchy of their society, and the result is often rebellion by these groups and repression by the state. Moreover, water developments also induce conflict among downstream users over water and irrigable land, as we saw in the Senegal River basin.[19]

Hypothesis Two: Environmental scarcity causes large population movement, which in turn causes group-identity conflicts.

There is substantial evidence to support this hypothesis. But we must be sensitive to "contextual factors" unique to each socio-ecological system. These are the system's particular physical, political, economic, and cultural features that affect the strength of the linkages between scarcity, population movement, and conflict.

For example, experts emphasize the importance of both "push" and "pull" factors in decisions of potential migrants. These factors help distinguish migrants from refugees: while migrants are motivated by a combination of push and pull, refugees are motivated mainly by push. Environmental scarcity is more likely to produce migrants than refugees, because it usually develops gradually, which means that the push effect is not sharp and sudden and that pull factors can therefore clearly enter into potential migrants' calculations.

Migrants are often weak and marginal in their home society, and, depending on context, they may remain weak in the receiving society. This limits their ability to organize and to make demands. States play a critical role here: migrants often need the backing of a state (either of the receiving society or an external one) before they have sufficient power to cause conflict, and this backing depends on the region's politics. Without it, migration is less likely to produce violence than silent misery and death, which is rarely destabilizing. We must remember too that migration does not always produce bad results. It can act as a "safety valve" by reducing conflict in the sending area. Depending on the economic context, it can ease labor shortages in the receiving society, as it sometimes has, for instance, in Malaysia. Countries as different as Canada, Thailand, and Malawi show the astonishing capacity of some societies to absorb migrants without conflict.[20]

Although we must recognize the role of such contextual factors, events in Bangladesh and Northeast India provide strong evidence in support of the second hypothesis. In recent decades, huge numbers of people have moved from Bangladesh to India, producing group-identity conflicts in the adjacent Indian states. Only one of the three sources of environmental scarcity—population growth—seems to be a main force behind this migration. Even though the Bangladesh cropland is heavily used, in general it is not badly degraded, because the annual flooding of the Ganges and Brahmaputra rivers deposits nutrients that help maintain the fertility of the country's floodplains. And while land distribution remains highly unequal, this distribution has changed little since an initial attempt at land-reform immediately following East Pakistan's independence from the British.[21]

But the United Nations predicts that Bangladesh's current population of 120 million will grow to 235 million by the year 2025.[22] At about 0.08 hectares per capita, cropland is already desperately scarce. Population density is over 900 people per square kilometer (in comparison, population density in neigboring Assam is under 300 per square kilometer). Since virtually all of the country's good agricultural land has been exploited, population growth will cut in half the amount of cropland available per capita by 2025. Land scarcity and the brutal poverty and social turmoil it engenders have been made worse by flooding (perhaps aggravated by deforestation in the Himalayan watersheds of the region's major rivers); the susceptibility of the country to cyclones; and the construction by India of the Farakka Barrage, a dam upstream on the Ganges River.

Of course, people have been moving around this part of South Asia in large numbers for centuries. But the movements are increasing in size. Over the last forty years, millions have migrated from East Pakistan or Bangladesh to the Indian states of Assam, Tripura, and West Bengal. Detailed data are scarce, since both India and Bangladesh manipulate their census data for political reasons. And the Bangladeshi government avoids admitting there is large out-migration, because the question causes friction with India. But our research pieced together demographic information and experts' estimates. We conclude that migrants from Bangladesh have expanded the population of neighboring areas of India by twelve to seventeen million, of which only one or two million can be attributed to migration induced by the 1971 war between India and Pakistan that created Bangladesh. We further estimate that the population of the state of Assam has been boosted by at least seven million people to a current total of twenty-two million.[23]

This enormous flux has produced pervasive social changes in the receiving regions. It has altered land distribution, economic relations, and the balance of political power between religious and ethnic groups, and it has triggered serious intergroup conflict. Members of the Lalung tribe in Assam, for instance, have long resented Bengali Muslim migrants: they accuse them of stealing the area's richest farmland. In early 1983, during a bitterly contested election for federal offices in the state, violence finally erupted. In the village of Nellie, Lalung tribespeople massacred nearly 1,700 Bengalis in one five-hour rampage.[24]

In Tripura, the original Buddhist and Christian inhabitants now make up less than 30 percent of the state's population. The rest consists of Hindu migrants from either East Pakistan or Bangladesh. This shift in the ethnic balance precipitated a violent insurgency between 1980 and 1988 that was called off only after the government agreed to return land to dispossessed Tripuris and stop the influx of

Bangladeshis. But, as the migration has continued, this agreement is in jeopardy.

There are important contextual features unique to this case. Within Bangladesh, key push factors include inheritance practices that divide cropland into smaller plots with each generation, and national and community water-control institutions that sharply limit agricultural output and keep peasants from gaining full benefit from some of the most fertile land in the world. On the "pull" side, the standard of living in India is markedly better, and Indian politicians have often encouraged Bangladeshi migration to garner votes. Furthermore, in the Ganges-Brahmaputra region, the concept of nation-state is often not part of the local culture. Many people think of the region as "greater Bengal," and borders do not figure heavily in the calculations of some migrants, especially when there are receptive family, linguistic, and religious groups across the frontier. Finally, during the colonial period, the British used Hindus from Calcutta to administer Assam, and Bengali became the official language. As a result, the Assamese are particularly sensitive to their loss of political and cultural control in the state.

While such contextual factors are important, they cannot obscure the fact that land scarcity in Bangladesh, arising largely from population growth, has been a powerful force behind migration to neighboring regions and communal conflict there.

Hypothesis Three: Environmental scarcity simultaneously increases economic deprivation and disrupts key social institutions, which in turn causes "deprivation" conflicts such as civil strife and insurgency.

Empirical evidence partially supports this third hypothesis. Environmental scarcity does produce economic deprivation, and this deprivation does cause civil strife. But more research is needed on the effects of scarcity on social institutions.

Resource degradation and depletion often affect economic productivity in poor countries and therefore contribute to deprivation. For example, Robert Repetto of the World Resources Institute estimates that erosion in upland Indonesia annually costs the country's agricultural economy nearly half a billion dollars in discounted future income.[25] The Magat watershed on the northern Filipino island of Luzon—a watershed representative of many in the country—suffers gross erosion rates averaging 219 tons per hectare per year; if the lost nutrients were replaced by fertilizer, the annual cost would be over $100 per hectare.[26] And dryland degradation in Burkina Faso reduces the country's GDP by nearly 9 percent because of fuelwood loss and lower yields of millet, sorghum, and livestock.[27]

I originally hypothesized that scarcity would undermine a variety of social institutions. Our research suggests, however, that one institution in particular—the state—is most important. Although more study is needed, the multiple effects of environmental scarcity, including large population movements and economic decline, appear likely to sharply weaken the capacity and legitimacy of the state in some poor countries.

First, environmental scarcity increases financial and political demands on governments. For example, to mitigate the social effects of water, soil, and forest loss, governments must spend huge sums on industry and infrastructure such as new dams, irrigation systems, fertilizer plants, and reforestation programs. Furthermore, this resource loss can reduce the incomes of elites directly dependent on resource extraction; these elites usually turn to the state for compensation. Scarcity also expands marginal groups that need help from government by producing rural poverty and by displacing people into cities where they demand food, shelter, transport, energy, and employment. In response to swelling urban populations, governments introduce subsidies that drain revenues, distort prices, and cause misallocations of capital, which in turn hinders economic productivity. Such large-scale state intervention in the marketplace can concentrate political and economic power in the hands of a small number of cronies and rent seekers at the expense of other elite segments and rural agricultural populations.

Second and simultaneously, if resource scarcity affects the economy's general productivity, revenues to local and national governments will decline. This hurts elites that benefit from state largesse and reduces the state's capacity to meet the increased demands arising from environmental scarcity. A widening gap between state capacity and demands on the state, along with the misguided economic interventions such a gap often provokes, aggravates popular and elite grievances, increases rivalry between elite factions, and erodes the state's legitimacy.

Key contextual factors affect whether lower economic productivity and state weakening lead to deprivation conflicts. Research on civil strife shows it to be a function of both the level of grievance motivating challenger groups and the opportunities available to these groups to act on their grievances. The likelihood of civil strife is greatest when multiple pressures at different levels in society interact to increase grievance and opportunity simultaneously. Our third hypothesis says that environmental scarcity will change both variables by contributing to economic crisis and by weakening institutions such as the state. But we recognized that numerous other factors also influence grievance and opportunity.

Contrary to common intuition, there is no clear correlation between poverty (or economic inequality) and social conflict. Whether or not people become aggrieved and violent when they find themselves increasingly poor depends, in part, on their notion of economic justice. For example, people belonging to a culture that inculcates fatalism about deprivation—as happens with lower castes in India—will not be as prone to violence as people believing they have a right to economic well-being. Theorists have addressed this problem by introducing the variable "relative deprivation."[28] But data show little correlation between measures of relative deprivation and civil conflict.[29]

Part of the problem is that analysts have commonly used aggregate behavioral data (such as GDP/capita and average educational levels) to measure individual deprivation. In addition, more recent research has shown that, to cause civil strife, economic crisis must be severe, persistent, and pervasive enough to erode the legitimacy, or moral authority, of the dominant social order and system of governance. System legitimacy is therefore a critical intervening variable between rising poverty and civil conflict. It is influenced by the aggrieved actors' subjective "blame system," which consists of their beliefs about who or what is responsible for their plight.

If serious civil strife is to occur, the structure of political opportunities facing challenger groups must keep them from effectively expressing their grievances peacefully but offer them openings for violence against authority. The balance of coercive power among social actors is key here, because it affects the probability of success and, therefore, the expected costs and benefits of different actions by the state, its supporters, and challenger groups. A state debilitated by corruption, by falling revenues and rising demand for services, or by factional conflicts within elites will be more vulnerable to violent challenges by political and military opponents; also vital to state strength is the cohesiveness of the armed forces and its loyalty to civil leadership.[30]

Challengers will have greater relative power if their grievances are articulated and actions coordinated through well-organized, well-financed, and autonomous opposition groups. Since grievances felt at the individual level are not automatically expressed at the group level, the probability of civil violence is higher if groups are already organized around clear social cleavages, such as ethnicity, religion, or class. These groups can provide a clear sense of identity and act as nuclei around which highly mobilized and angry elements of the population, such as unemployed and urbanized young men, quickly coalesce. Of course, to the extent that economic crisis weakens challenger groups more that the state, or affects mainly disorganized people, it will not lead to violence.

Some contextual factors can influence both grievance and opportunity, including the leadership and ideology of challenger groups, and international shocks and pressures, such as changes in trade and debt relations and in costs of imported factors of production like energy. The rapid growth of urban areas in poor countries may have a similar dual effect: people concentrated in slums can communicate more easily than those in scattered rural villages; this may reinforce grievances and, by reducing problems of coordination, also increase the power of challenger groups. Research shows, however, surprisingly little historical correlation between rapid urbanization and civil strife;[31] and the exploding cities of the developing world have been remarkably quiescent in recent decades. This may be changing: India has lately witnessed ferocious urban violence, often in the poorest slums, and sometimes directed at new migrants from the countryside.[32] Fundamentalist opposition to the Egyptian government is also located in some of the most desperate sectors of Cairo and other cities like Asyut.

The Philippines provides evidence of the links between environmental scarcity, economic deprivation, and civil strife. The country has suffered from serious strife for many decades, usually motivated by economic stress.[33] Today, cropland and forest degradation in the uplands sharply exacerbates this economic crisis. The current upland insurgency—including guerrilla attacks and assaults on military stations—is motivated by the poverty of landless agricultural laborers and farmers displaced into the remote hills, where the central government is weak.[34] During the 1970s and 1980s, the Communist New People's Army and the National Democratic Front found upland peasants receptive to revolutionary ideology, especially where coercive landlords and local governments left them little choice but to rebel or starve. The insurgency has waned somewhat since President Marcos left, not because economic conditions have improved much in the countryside but because the democratically elected central government is more legitimate and the insurgent leadership is ideologically rigid.

Contextual factors are key to a full understanding of this case. For instance, property rights governing upland areas are, for the most part, either nonexistent or very unclear. Legally these areas are a public resource and their "open access" character encourages inmigration. Yet many upland peasants find themselves under the authority of concessionaires and absentee landlords who have claimed the land. Neither peasants, concessionaires, nor landlords, though, have secure enough title to have incentive to protect the land. The country's external debt is another important factor, as it is with many poor countries. Rising debts encouraged the Marcos government, under pressure from international financial agencies, to adopt draconian stabilization and structural adjustment policies. This caused an economic crisis in the first half of the 1980s, which boosted agricultural unemployment, reduced opportunities for alternative employment in urban and rural industries, and gave a further push to migration into the uplands.[35]

Causal processes like those in the Philippines can be seen around the planet. Population growth and unequal access to good land force huge numbers of rural people into cities or onto marginal lands. In the latter case, they cause environmental damage and become chronically poor. Eventually these people may be the source of persistent upheaval, or they may migrate yet again, stimulating ethnic conflicts or urban unrest elsewhere.

A Combined Model

There are important links between the processes identified in the second and third hypotheses. For example, although population movement is sometimes caused directly by scarcity, more often it arises from the greater poverty caused by this scarcity. Similarly, state weakening increases the likelihood not only of deprivation conflicts, but of group-identity conflicts too.

It is useful, therefore, to bring the hypotheses together into one model of environment-conflict linkages, as in figure 3.2. Decreases in the quality and quantity of renewable resources, population growth, and unequal resource access act singly or in various combinations to increase the scarcity for certain groups of cropland, water, forests, and fish. This can reduce economic productivity, both for the local groups experiencing the scarcity and the larger regional and national economies. The affected people may migrate or be expelled to new lands. Migrating groups often trigger ethnic conflicts when they move to

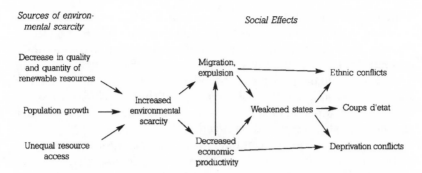

Figure 3.2 Some Sources and Consequences of Environmental Scarcity

new areas, while decreases in wealth can cause deprivation conflicts, such as insurgency and rural rebellion. The migrations and productivity losses may eventually weaken the state in developing countries, which in turn decreases central control over ethnic rivalries and increases opportunities for insurgents and elites challenging state authority. Figure 3.3 shows how these linkages work in the Filipino case.

South Africa and Haiti illustrate this combined model. In South Africa, apartheid concentrated millions of blacks in some of the country's least productive and most ecologically sensitive territories, where population densities were worsened by high natural birth rates. In 1980, rural areas of the Ciskei homeland had a population density of eighty-two per square kilometer, whereas the surrounding

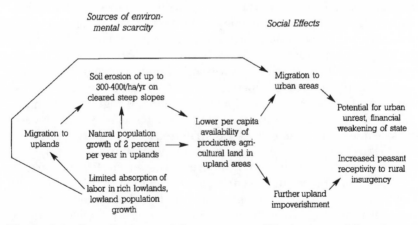

Figure 3.3 Some Sources and Consequences of Environmental Scarcity in the Philippines

Cape Province had a rural density of two. Homeland residents have little capital and few resource-management skills and are the victims of corrupt and abusive local governments. Sustainable development in such a situation is impossible, and wide areas have been completely stripped of trees for fuelwood, grazed down to bare dirt, and eroded of top soil. A 1980 report concluded that nearly 50 percent of Ciskei's land was moderately or severely eroded, and nearly 40 percent of its pasturage was overgrazed.[36]

This loss of resources, combined with a lack of alternative employment and the social trauma caused by apartheid, has created a subsistence crisis in the homelands. Thousands of people have migrated to South African cities, which are as yet incapable of adequately integrating and employing these migrants, due to the social and economic structures left by apartheid. The result is the rapid growth of squatter settlements and illegal townships that are rife with discord and that threaten the country's move to democratic stability.[37]

In Haiti, the irreversible loss of forests and soil in rural areas deepens an economic crisis that spawns social strife, internal migration, and an exodus of boat people. When first colonized by the Spanish in the late fifteenth century and the French in the seventeenth century, Haiti was treasured for its abundant forests. Since then, Haiti has experienced one of the world's most dramatic examples of environmental despoliation. Less than 2 percent of the country remains forested, and the last timber is being felled at 4 percent per year.[38] As trees disappear, erosion follows, worsened by the steepness of the land and harsh storms. The United Nations estimates that at least 50 percent of the country is affected by topsoil loss that leaves the land "unreclaimable at the farm level."[39] So much soil washes off the slopes that the streets of the capital, Port-au-Prince, have to be cleared with bulldozers in the rainy season.

Unequal land distribution was not a main cause of this catastrophe. Haiti gained independence in 1804 following a revolt of slaves and exslaves against the French colonial regime. Over a period of decades, the old plantation system, associated with slavery, was dismantled and land widely distributed in small parcels. As a result, Haiti's agricultural structure is unique to Latin America, with 73 percent of cropland in private farms of less than four hectares.[40]

But inheritance customs and population growth have combined to produce scarcity, as in Bangladesh. Land has been subdivided into smaller portions with each generation. Eventually the plots cannot properly support their cultivators, fallow periods are neglected, and greater poverty prevents investment in soil conservation. The poorest leave for steeper hillsides, where they clear the forest and begin farming anew, only to exhaust the land again in a few years. Many peasants try to supplement their falling incomes by scavenging wood for charcoal production, which contributes to further deforestation.

These processes might have been prevented had a stable central government invested in agriculture, industrial development, and reforestation. Instead, since independence Haiti has endured a ceaseless struggle for power between black and mulatto classes, and the ruling regimes have been solely interested in expropriating any surplus wealth the economy generated. Today, over 60 percent of the population is still engaged in agriculture, yet capital is unavailable for agricultural improvement, and the terms of exchange for crop production favor urban regions.[41] The population growth rate has actually increased, from 1.7 percent in the mid-1970s to over 2 percent today: the United Nations estimates that the current population of 6.75 million will grow to over 13 million by 2025.[42] As the land erodes and the population grows, incomes shrink: agricultural output per capita has decreased 10 percent in the last decade.[43]

Analysts agree that rising rural poverty has caused ever-increasing rural-rural and rural-urban migration. In search of work, agricultural workers move from subsistence hillside farms to rice farms in the valleys. From there, they go to cities, especially to Port-au-Prince, which now has a population of over a million. Wealthier farmers and traders, and eventually even those with slimmer resources, try to flee by boat.

These economic and migration stresses are undoubtedly contributing to civil strife. In the aftermath of the collapse of the "Baby Doc" Duvalier regime in 1986, the poor vented their vengeance on those associated with the regime, in particular on Duvalier's informal gangs of enforcers, collectively called *tontons macoutes*. During his election campaign and his short tenure as president, Jean-Bertrand Aristide reportedly encouraged poor slum dwellers to attack Haiti's elite. Fearful of uprisings, the military regime ferociously oppressed the country's poor and peasantry. Haiti will forever bear the burden of its irreversibly ravished environment, which may make it impossible to build a prosperous, just, and peaceful society.

The Causal Role of Environmental Scarcity

Debate about whether and how environmental scarcity contributes to conflict often centers on the specific causal role of scarcity. Here I discuss the following variables that characterize this role: necessity, strength, proximity, exogeneity, multicausality, nonlinearity and interactivity.

Necessity is a dichotomous variable: something is either a necessary cause of a given type of event, or it is not. Environmental scarcity is not a necessary cause of violent conflict, since much violence occurs in situations of resource abundance. Unlike necessity, the *strength* of a cause can be measured along a continuum, from weak to "sufficient." Causal *proximity* can similarly be measured along a continuum, from distant to proximate. We commonly think of proximity in terms of causal distance in time or space. But proximity is really a function of the number of intervening causal steps or variables between the cause and its effect; the larger the number of intervening variables, the lower the causal proximity. The characteristics of proximity and causal strength are sometimes conflated, since a distant cause is often assumed to be weak. But intervening variables do not necessarily weaken the link between a cause and its effect. In the cases described in this chapter, environmental scarcity was not a proximate cause of conflict, but it played a powerful underlying role nonetheless.

The causal independence of a variable, or its *exogeneity*, can also be measured along a continuum, from fully endogenous to fully exogenous. Many analysts assume that environmental scarcity is no more than a fully endogenous intervening variable linking political, economic, and social factors to conflict. By this view, environmental scarcity may be an important indicator that political and economic development has gone awry, but it does not merit, in and of itself, intensive research and policy attention. Instead, we should devote our resources to the more fundamental political and economic factors.

But the cases reviewed here highlight three reasons why this view is not entirely correct. First, as we have seen in the Senegal basin, environmental scarcity can be an important force behind changes in the politics and economics governing resource use. Scarcity caused a powerful actor to strengthen in its favor an inequitable distribution of resources. Second, ecosystem vulnerability is often an important variable contributing to environmental scarcity, and this vulnerability is, at least in part, a physical given. The depth of soils in the Filipino uplands is not a function of human social institutions or behavior. Third, in many parts of the world—including regions of the Philippines, China, India, Haiti, and South Africa—environmental degradation has crossed a threshold of irreversibility. Even if enlightened social change removes the original political, economic, and cultural causes of the degradation, it will be a continuing burden on society. Once irreversible, in other words, environmental degradation becomes an exogenous variable.

The degree of *multicausality* of the processes producing social conflict also varies. As we have seen, if environmental scarcity contributes to conflict, it almost always operates with other political, economic, and cultural factors. Once again, though, skeptics often conflate the characteristics of multicausality and causal strength by assuming that if many factors are involved, each must be relatively weak. But our cases show that environmental change does not have to be the sole cause of conflict, or even one of only a few causes, to be a strong one.

Interactivity is another dichotomous variable: the relationship among multiple causes of an event can be either interactive or additive. Interaction is a common feature of environmental-social systems. In an interactive system, none of the causes is sufficient but all are necessary; thus causal strength and interactivity are linked because no single cause can produce the event itself. But beyond this statement, it is meaningless to claim that a given cause in an interactive system is stronger than another. Finally, the degree of *nonlinearity* of the individual mathematical functions describing the relations between a system's elements can vary from high to low. A system with highly nonlinear functions will probably exhibit unanticipated "threshold effects" and chaotic behavior in response to small perturbations. This is a key characteristic of some environmental-social systems.

Academic and lay discussions of environment-conflict linkages are usually larded with imprecise causal verbs like *aggravate, amplify,* and *trigger.* These fuzzy, folk concepts are useful in everyday explanations of physical and social events, but they are not always helpful for research. We can clarify them a bit, however, using the above distinctions.

A claim that an environmental factor "amplifies" the effect of other causes of conflict implies that the factor *interacts* with the other causes to multiply their impact. In contrast, if the factor "aggravates" the impact of the other causes, this suggests the factor's effect is *added* to that of the others. A "trigger" of conflict is always a proximate cause, and usually an unnecessary and insufficient one too (as was the assassination of Archduke Ferdinand as a cause of World War I). The term also implies that the system responds nonlinearly to the factor in question; that is, the factor triggers a disproportionately large response by pushing the system beyond a critical threshold. Stochastic and extreme environmental events—such as cyclones, floods, and droughts—can be important triggers of conflict. They can provide challenger groups with opportunities for action against a state whose buffering capacity has been gradually eroded by civil war, corruption, economic mismanagement, rapid population growth, or deteriorating stocks of renewable resources.[44]

Aggravator, amplifier, and *trigger* models are popular with skeptics because they relegate environmental factors to the status of secondary variables. While these models are often valuable, they offer inaccurate and incomplete explanations of important cases. Environmental stresses can be important contributors to conflict even if causally distant and even if the system is complex and highly interactive.

Implications for International Security

Environmental scarcity has insidious and cumulative social impacts, like population movement, economic decline, and the weakening of states. These can contribute to diffuse, persistent, and subnational violence. The rate and extent of such conflicts will increase as scarcities worsen.

This subnational violence will not be as conspicuous or dramatic as interstate resource wars, but it will have serious repercussions for the security interests of both the developed and developing worlds. Countries under such stress may fragment as their states become enfeebled and peripheral regions are seized by renegade authorities and warlords. Governments of countries as different as the Philippines and Peru have lost control over outer territories; although both these cases are complicated, environmental stress has contributed to fragmentation. Fragmentation of any sizeable country will produce large outflows of refugees; it will also prevent the country from effectively negotiating and implementing international agreements on collective security, global environmental protection, and other matters.

A state might keep scarcity-induced civil strife from causing its progressive enfeeblement and fragmentation by becoming a "hard" regime that is authoritarian, intolerant of opposition, and militarized. Such regimes are more prone to launch military attacks against neighboring countries to divert attention from internal grievances. If a number of developing countries evolve in this direction, they could eventually threaten the military and economic interests of rich countries.

A state's ability to follow the second path in the response to environmentally induced turmoil depends, I believe, on two factors. First, the state must have sufficient remaining capacity—despite the debilitating effects of scarcity—to mobilize or seize resources for its own ends; this is a function of the internal organizational coherence of the state and its autonomy from outside pressures. Second, there must remain enough surplus wealth in the country's ecological-economic system to allow the state, once it seizes this wealth, to pursue its authoritarian course. Consequently, the countries with the highest probability of becoming "hard" regimes, and potential threats to their neighbors, are large, relatively wealthy developing countries that are dependent on a declining environmental base and that have a history of state strength. Candidates include Indonesia, Brazil and, perhaps, Nigeria.

Conclusions

Our research shows that environmental scarcity can contribute to violent conflict. This conflict tends to be persistent, diffuse, and subnational. Its frequency will probably jump sharply in the next decades as scarcities rapidly worsen in many parts of the world. Of immediate concern are scarcities of cropland, water, forests, and fish. Atmospheric changes such as global warming will probably not have a major effect for several decades, and then mainly by interacting with already existing scarcities.

The degradation and depletion of environmental resources is only one source of environmental scarcity; two other important sources are population growth and unequal resource distribution. Scarcity often has its harshest social impact when these factors interact. As environmental scarcity becomes more severe, some societies will have a progressively lower capacity to adapt. Of particular concern here is the decreasing capacity of the state to create markets and other institutions that promote adaptation. The impact of environmental scarcity on state capacity therefore deserves further research.

Countries experiencing chronic internal conflict because of environmental stress will probably either fragment or become more authoritarian. Either outcome could seriously disturb international relations. Authoritarian regimes may be inclined to launch attacks against other countries to divert popular attention from internal stresses. Fragmenting countries will be the source of large out-migrations, and they will be unable to effectively negotiate or implement international agreements on security, trade, and environmental protection.

—————————————— NOTES ——————————————

This chapter is an edited version of "Environmental Scarcities and Violent Conflict: Evidence from Cases," in *International Security* vol. 19, no. 1 (1994). The author thanks the participants in the project on Environmental Change and Acute Conflict, especially project codirectors Jeffrey Boutwell and George Rathjens.

1. Thomas Homer-Dixon, "On the Threshold: Environmental Changes As Causes of Acute Conflict," *International Security* vol 16, no. 2 (Fall 1991), pp. 76–116.

2. The three-year project on Environmental Change and Acute Conflict brought together a team of thirty researchers from ten countries. It was sponsored by the American Academy of Arts and Sciences and the Peace and Conflict Studies Program at the University of Toronto.

3. On simple-scarcity conflicts, we examined water in the Jordan and Nile river basins and the Southern African region; on environmentally induced group-identity conflicts, we focused on Bangladesh-Assam and the Miskito Indians in Nicaragua; and on economic decline and civil strife, we studied the Philippines and China. Researchers within the project also investigated the 1989 conflict in the Senegal River basin, the 1969 Soccer War between El Salvador and Honduras, the rise of the Sendero Luminoso in Peru, migration and civil strife in Haiti, and migration from black homelands in South Africa.

4. Diana Liverman, "The Impacts of Global Warming in Mexico: Uncertainty, Vulnerability and Response," in Jurgen Schmandt and Judith Clarkson, eds., *The Regions and Global Warming: Impacts and Response Strategies* (New York: Oxford University Press, 1992), pp. 44–68; and Diana Liverman and Karen O'Brien, "Global Warming and Climate Change in Mexico," *Global Environmental Change* vol. 1, no. 4 (December 1991), pp. 351–64.

5. The second and third types of scarcity arise only with resources that can be physically controlled and possessed, like fish, fertile land, trees, and water, but unlike the climate or the ozone layer.

6. Since population growth is often a main cause of a decline in the quality

and quantity of renewable resources, it actually has a dual impact on resource scarcity, a fact rarely noted by analysts.

7. Jeffrey Leonard, "Overview," *Environment and the Poor: Development Strategies for a Common Agenda* (New Brunswick, N.J.: Transaction, 1989), p. 7. For a careful analysis of the interaction of population and land distribution in El Salvador, see chapter 2 in William Durham, *Scarcity and Survival in Central America: The Ecological Origins of the Soccer War* (Stanford, Calif.: Stanford University Press, 1979), pp. 21–62.

8. Global Assessment of Soil Degradation, *World Map on Status of Human-Induced Soil Degradation*, Sheet 2, Europe, Africa, and Western Asia (United Nations Environment Programme; International Soil Reference Centre, Wageningen, the Netherlands, 1990).

9. Nafis Sadik, *The State of the World Population 1991* (New York: United Nations Population Fund, 1991), p. 24; World Resources Institute, *World Resources 1992–93* (New York: Oxford University Press, 1992), pp. 246, 262.

10. G. M. Higgins, et al., *Potential Population Supporting Capacities of Lands in the Developing World*, Technical Report of Project INT/75/P13, "Land Resources of the Future," undertaken by the Food and Agriculture Organization of the United Nations in collaboration with the International Institute for Applied Systems Analysis and the United Nations Fund for Population Activities (Rome, 1982), p. 137.

11. Michael Horowitz, "Victims of Development," *Development Anthropology Network*, Bulletin of the Institute for Development Anthropology vol. 7., no. 2 (Fall 1989), pp. 1–8; and Horowitz, "Victims Upstream and Down," *Journal of Refugee Studies* vol. 4, no. 2 (1991), pp. 164–81.

12. A full account can be found in Maria Concepcion Cruz et al., *Population Growth, Poverty, and Environmental Stress: Frontier Migration in the Philippines and Costa Rica* (Washington, D.C.: World Resources Institute, 1992).

13. World Bank, *Philippines: Environment and Natural Resource Management Study* (Washington, D.C.: World Bank, 1989). Erosion rates can exceed three hundred tons per hectare per year, which is ten to twenty times the sustainable rate.

14. Gareth Porter and Delfin Ganapin, Jr., *Resources, Population, and the Philippines' Future: A Case Study*, WRI Paper No. 4 (Washington, D.C.: World Resources Institute, 1988).

15. Arthur Westing, "Appendix 2. Wars and Skirmishes Involving Natural Resources: A Selection from the Twentieth Century," in Arthur Westing, ed., *Global Resources and International Conflict: Environmental Factors*

in Strategic Policy and Action (New York: Oxford University Press, 1986), pp. 204–10.

16. See Durham, *Scarcity and Survival*.

17. Peter Gleick, "Water and Conflict," Occasional Paper No. 1, Project on Environmental Change and Acute Conflict (September 1992); and "Water and Conflict: Fresh Water Resources and International Security," *International Security*.

18. "Pretoria Has Its Way in Lesotho," *Africa Report* (March–April, 1986), pp. 50–51; Patrick Laurence, "A 'New Lesotho'?" *Africa Report* (January–February 1987), pp. 61–64; "Lesotho Water Project Gets Under Way," *Africa Report* (May–June 1988), p. 10. See also Charles Okidi, "Environmental Stress and Conflicts in Africa: Case Studies of African International Drainage Basins," paper prepared for the Project on Environmental Change and Acute Conflict (May 1992).

19. See Thayer Scudder, "River Basin Projects in Africa," *Environment* vol. 31, no. 2 (March 1989), pp. 4–32; and Scudder, "Victims of Development Revisited: The Political Costs of River Basin Development," *Development Anthropology Network*, Bulletin of the Institute for Development Anthropology vol. 8, no. 1 (Spring 1990), pp. 1–5.

20. Astri Suhrke, "Pressure Points: Environmental Degradation, Migration, and Conflict," Occasional Paper No. 3, Project on Environmental Change and Acute Conflict (March 1993).

21. James Boyce, *Agrarian Impasse in Bengal: Institutional Constraints to Technological Change* (Oxford, UK: Oxford University Press, 1987), p. 9.

22. Nafis Sadik, *The State of the World Population 1991*, p. 43.

23. Sanjoy Hazarika, "Bangladesh and Assam: Land Pressures, Migration, and Ethnic Conflict," Occasional Paper No. 3, Project on Environmental Change and Acute Conflict (March 1993), p. 52–54.

24. "A State Ravaged," *India Today*, March 15, 1983, pp. 16–21; "Spillover Tension," *India Today*, March 15, 1983, pp. 22–23. The 1991 Indian Census showed that Assam's population growth rate has declined; the conflicts in Assam in the early 1980s appear to have encouraged many migrants from Bangladesh to go to West Bengal instead.

25. Robert Repetto, "Balance-Sheet Erosion—How to Account for the Loss of Natural Resources," *International Environmental Affairs* vol. 1, no. 2 (Spring 1989), pp. 103–37.

26. This estimate does not include the economic costs of lost rooting depth and increased vulnerability to drought, which may be even larger. See

Wilfrido Cruz, Herminia Francisco, and Zenaida Conway, "The On-Site and Downstream Costs of Soil Erosion in the Magat and Pantabangan Watersheds," *Journal of Philippine Development* vol. 15, no. 1 (1988), p. 88.

27. Ed Barbier, "Environmental Degradation in the Third World," in David Pearce, ed., *Blueprint 2: Greening the World Economy* (London: Earthscan, 1991), Box 6.8, p. 90.

28. People are relatively deprived when they perceive a widening gap between the level of satisfaction they have achieved (usually defined in economic terms) and the level they believe they deserve. Deprivation is therefore *relative to* some subjective standard of equity or fairness, and the size of the perceived gap obviously depends on the beliefs about economic justice held by the individual. See Ted Gurr, *Why Men Rebel*, (Princeton, N.J.: (Princeton University Press, 1970).

29. Steven Finkel and James Rule, "Relative Deprivation and Related Theories of Civil Violence: A Critical Review," in Kurt and Gladys Lang, eds., *Research in Social Movements, Conflicts, and Change* (Greenwich, Conn.: JAI, 1986), pp. 47–69.

30. See Farrokh Moshiri, "Revolutionary Conflict Theory in an Evolutionary Perspective," and Jack Goldstone, "An Analytical Framework," both in Jack Goldstone, Ted Gurr, and Farrokh Moshiri, eds., *Revolutions of the Late Twentieth Century* (Boulder, Colo.: Westview, 1991), pp. 4–36, 37–51.

31. Wayne Cornelius, Jr., "Urbanization as an Agent in Latin American Political Instability: The Case of Mexico," *American Political Science Review* vol. 63 (1969), pp. 833–57; and Abdul Lodhi and Charles Tilly, "Urbanization, Crime, and Collective Violence in Nineteenth-Century France," *American Journal of Sociology* vol. 79, no. 2 (September 1973), pp. 296–318.

32. Sanjoy Hazarika, "Week of Rioting Leaves Streets of Bombay Empty," *New York Times*, January 12, 1993, p. A3.

33. The Huk rebellion in the late 1940s and early 1950s provides some of the best evidence for the link between economic conditions (especially unequal land distribution) and Filipino civil strife. See Benedict Kerkvliet, *The Huk Rebellion: A Study of Peasant Revolt in the Philippines* (Quezon City, Philippines: New Day Publishers, 1979); and E. J. Mitchell, "Some Econometrics of the Huk Rebellion," *American Political Science Review* vol. 63, no. 4 (December 1969), pp. 1159–71.

34. Celso Roque and Maria Garcia, "Economic Inequality, Environmental Degradation and Civil Strife in the Philippines," paper prepared for the Project on Environmental Change and Acute Conflict (1993).

35. Maria Concepcion Cruz and Robert Repetto, *The Environmental*

Effects of Stabilization and Structural Adjustment Programs: The Philippines Case (Washington, D.C.: World Resources Institute, 1992).

36. Francis Wilson and Mamphela Ramphele, *Uprooting Poverty: The South African Challenge* (New York: Norton, 1989); George Quail et al., *Report of the Ciskei Commission* (Pretoria: Conference Associates, 1980), p. 73.

37. See Mamphela Ramphele and Chris McDowell, eds., *Restoring the Land: Environment and Change in Post-Apartheid South Africa* (London: Panos, 1991); and Chris Eaton, "Rural Environmental Degradation and Urban Conflict in South Africa," Occasional Paper of the Peace and Conflict Studies Program, University of Toronto, June, 1992.

38. *World Resources, 1992–93*, p. 286.

39. Global Assessment of Soil Degradation, *World Map on Status of Human-Induced Soil Degradation*, Sheet 1, North and South America.

40. Anthony Catanese, "Haiti's Refugees: Political, Economic, Environmental," *Field Staff Reports*, No. 17 (Sausalito, Calif.: Universities Field Staff International, Natural Heritage Institute, 1990–91), p. 5.

41. Marko Ehrlich et al., *Haiti: Country Environmental Profile, A Field Study* (Washington, D.C.: USAID, 1986), pp. 89–92.

42. *World Resources, 1992–93*, p. 246.

43. Ibid., p. 272.

44. Such opportunistic exploitation of drought in Africa is discussed by Peter Wallensteen in "Food Crops as a Factor in Strategic Policy and Action," in Westing, ed., *Global Resources*, pp. 154–55.

4

A Realist's Conceptual Definition of Environmental Security

Michel Frédérick

I n the field of international relations, analysts favoring the realist
paradigm agree on a number of basics. One, the international
system is fundamentally anarchic in the sense that political life
therein is dominated by states not subject to any authority superior to
their own sovereignty. Two, within this anarchic system, states will
relentlessly pursue their respective national interests in order to ensure their survival. Three, the core concept of national interest is to be
defined in terms of power as the only means of attaining national
prosperity and security. Four, foreign policies will consequently seek

either to keep power, to increase power, or to demonstrate power. Five, within the anarchic system, states are in constant competition which, however, does not prevent them from accomodating their respective interests through cooperation when it directly serves them in their quest for national prosperity and security.

The New World Order dreamed about by many idealists as a result of the significant events occurring from 1989 to 1991 and expected to bring about an end to the anarchy of the international system should have tolled the bell for the realist paradigm especially as it was laying bare some of its weaknesses and contradictions, like the fact that it had been unable to foretell or clearly explain the outcome of the Cold War and the collapse of the Soviet Empire.[1] Today, however, it appears clear that this New World Order will not come to pass. The unexpected failure of the Somalia intervention, the disaster in the former Yugoslavia, the typical hesitation waltz that occurred over Haiti or again Rwanda, and the plight of refugees in Zaire, daily call to mind the image of an order—or an anarchy—believed to have vanished but which obviously still exists even though presented under different appearances. This can also be seen in the outcomes of major diplomatic conferences, such as the one in Rio in 1992 on the environment and the one in Cairo in 1994 on population, both anticipating a desire for closer relations between the North and the South along with the pooling of concrete development measures. In the final analysis, these conferences provided little to underdeveloped countries whose concerns still remain clearly subordinate to the dominating economies of the North.

On the other hand, it is too soon as yet to begin talking about a New World Disorder. Despite the fact that the global agenda was shaken up—an end to bipolarity, the resurgence of regionalism, a return to economic continentalism, the evolution of international institutions and regimes, the surfacing of auxiliary political stakes such as the environment, human rights, and the protection of minorities—when it comes to the bedrock of the international system it remains essentially unchanged. An endless pursuit of national self-interest actually remains a dominant stake. Moreover, the rush of the former Soviet "colonies" to attain sovereignty and statehood managed to confound all those experts who, just a few years earlier, had pronounced the concept of the sovereign nation-state obsolete and lacking any future.

For realist scholars, this evolution is not without meaning. The pursuit of national self-interest encourages states to interpret many situations in terms of national security. Realizing that security policies have always had a double goal, namely to preserve the integrity of the territory of the state as well as its sovereignty, we can see how, in a mostly interstate system, the policy of international security came to reflect this concern and how it too came to revolve around these two interdependent notions.[2] Now, as this international security policy proceeds to embody other aspects—such as the environment—and to grant them significant importance, it brings about for researchers the task of delimiting, within the context of the pursuit of national self-interest, both the meaning and the reach of the new concepts emerging from it. Therefore, this text has assumed the objective of contributing elements that can define, from a realist perspective, the meaning and reach of the environmental security concept.

I. The Emergence of the Environmental Security Concept

Over recent decades, events such as the oil crises of 1973 and 1979, the protectionist policies of certain trading powers, and the widening of inequalities between the North and the South underlined the impact of economic factors when formulating questions about security. No longer could they be asked in the same terms once we began referring to "economic security" in the same way as we came to speak about "social security" a few decades earlier when states found themselves threatened from within by unrest sparked by an unequal distribution of the collective wealth to the detriment of the unemployed, the ill, and the aged.

Strategic analysis experts thus had to face the truth and acknowledge that the "nonmilitary dimensions of security would henceforth occupy a very important place in state behaviorism."[3] This led them to redefine the concept of national security so as to broaden its scope,[4] as the traditional approach reflected too narrow a vision not only of the problems but also of the solutions available to cope with threats.[5] It became essential to come up with a broader definition that would take into consideration the fact that a threat to national security exists once an action or sequence of events "threatens . . . to degrade the quality of life for the inhabitants of a state or . . . threatens significantly to narrow the range of policy choices available to the government of a state or to private nongovernmental entities within the state."[6]

It was from within the framework of discussions surrounding these redefining efforts that the concept of environmental security surfaced.[7] Actually, this emergence became inevitable once "national security" became associated with "quality of life" within a sociopolitical context marked over recent years by the introduction of environmental questions into overall national and international concerns. There were two determining factors involved in the emergence of the concept: the growing shrillness of the environmental threat and the unreliability of international mechanisms for managing the environment.

A. The Environmental Threat

In the early 1960s nobody was talking about the environment. There were virtually no international agreements nor even any national laws in this area. No country had as yet set up a real department of the environment and the ecology movement had barely taken root. The situation, however, was to evolve rapidly toward what could be called a rediscovery, within industrialized societies, of close interdependence between man and nature. Works of note such as *Silent Spring* and *Limits to Growth* drew attention to the depletable aspect of the natural and energy resources required for our modern society to function, as well as to the environmental threat triggered by uncontrolled industrial growth.

In the United States, events such as Love Canal and Three Mile Island roused public opinion: as marginal as they were ten years earlier, environmental problems were henceforth to be on the agenda of political decisionmakers. Their rise became unstoppable and even incited President Carter in 1977 to set up a major task force mandated to draw up a prospective report on an environmental crisis which could have a profound impact on the planet between now and the year 2000.[8] At the same time, both in the United States and Canada as well as in Continental Europe and Scandinavia, an awareness of the gravity of the acid rain phenomenon began to sink in. Taking into account the frequently transborder source of this type of pollution, the problem of acid rain underlined, more than ever before, the need for increased interstate dialogue in seeking a lasting solution. But it also led states to view more clearly the problems of transborder pollution as an invasion of their territorial integrity and hence a threat to their national security.

This trend became even more pronounced during the 1980s as environmental problems became more significant in terms of numbers and especially in terms of complexity. Indeed, global phenomena whose mechanics and repercussions present complex interactions—I am referring here to the greenhouse effect and its ensuing climatic changes, to the thinning of the ozone layer, and to the loss of biological diversity—were added to an already lengthy list of problems (desertification, deforestation, marine pollution, radioactive waste, industrial and household waste, erosion of tillable soil, etc.) which, especially if taken collectively, constitute a very real threat to humans and their institutions. And if we add to this already serious situation the three sociopolitical causes of environmental deterioration—namely, poverty, underdevelopment and the squandering of human and financial resources—we can better comprehend, when faced with social unrest and conflicts that could result, why more and more states are treating environmental problems as national security matters.

This growth of the environmental threat was well documented in the Brundtland Report (1987) and in the preparatory papers for the Rio de Janeiro Summit (1992). While these explore certain remedial avenues essentially founded on sustainable development, they also illustrate a deficiency in the existing international system, which constitutes a second factor behind the emergence of the environmental security concept. It involves the unreliability, at an international level, of the mechanisms for managing the environment.

B. The Unreliability of International Mechanisms for Managing the Environment

Each year tens of millions of tons of waste and polluting substances are cast directly into the sea despite the fact that such practices are prohibited and strictly regulated by over twenty international or regional agreements on the prevention of marine pollution. Among many others, this example clearly underlines the unreliability of the mechanisms for managing the environment on an international scale.

Even though for some twenty years people have usually ranked quality of the environment as a priority concern for the international community, this recognition still has not engendered a genuine right to a safeguarded environment.[9] The inevitable result of this situation is that states and international institutions are led to reassess the role international law can play in protecting the planet's environment in view of the difficulties they must face in adapting its norms to the ever-increasing pace of environmental changes. Moreover, the ineffectiveness of many current norms is responsible for creating among them a certain skepticism when it comes to the very ability of international law to translate the contents of international public policies into concrete and applicable norms.

We cannot talk about the unreliability of existing international mechanisms without also discussing the input of international organizations dealing with the management of environmental problems. Indeed, the consultation and decision-making processes involving environmental matters are still very poorly organized at the international level. Despite the obvious links between the environment, the economy and international development, organizations specializing in the two latter fields—coming to mind are the World Bank, the United Nations Food and Agriculture Organization (FAO), and the Organization for Economic Cooperation and Development (OECD)—are not yet equipped with sufficient resources to enable them to intervene more effectively, especially in a more concerted way, to resolve global environment problems. As for the United Nations Environment Program (UNEP), should it wish to broaden its role, its status of simple "program" prevents it from doing so: it has only the power of recommendation. Despite this, its initiatives have been numerous, most notably on the level of normalcy. However, it remains obvious, as illustrated by its setbacks, that it will not be able to break out of the framework of established solutions until its mandate is beefed up and it acquires real powers of intervention.[10]

This unreliability of the existing mechanisms for managing the environment introduces an element of uncertainty into the international system, leading states themselves to keep a close eye on their interests in the environmental field. Realism would postulate in this regard that states will invariably seek security against the perils of nature and of environmental degradation and change. Consequently, it is through the prism of their own national security, and hence their own national interest, that states will analyze the impact of global environment problems along with state behavior apt to upgrade the situation or further harm it. Within such a context, the environmental security concept will then take on all the importance that states will effectively grant it.

II. The Scope of the Concept of Environmental Security

What empirical reality do we want to describe with the expression "environmental security"? Political scientists who have tackled the question have all done so in case studies dealing with the consequences of the degradation of ecosystems and the depletion of natural resources. Two trends emerge: the first is focused on the environment and the second on the state.

A. Security of the Environment

In studies favoring this perspective the environment is considered a dependent variable: environmental security thus takes on the meaning of security of the environment.[11]

Security of the environment involves three elements: (1) the sustainable use of renewable and nonrenewable resources; (2) protection of the elements—air, water, soil—so as to prevent pollution from stifling natural regeneration; and (3) the maximum reduction of hazards related to industrial activities.[12] According to Arthur Westing, its scope covers the overall problems associated with the protection and use of the environment.[13] Such security could be compromised by acts of vandalism (war and ecoterrorism), excessive pollution, or the nonsustainable use of resources along with permanent human intrusion into certain ecologically sensitive areas.

This idea of environmental security stems from an essentially globalist view of interstate relations. Actually, it considers the security of states in only a minor way; its main concern is the security of the planet taken in its totality. In studies favoring this view, much interest is shown in the impact of global environmental problems such as the greenhouse effect, the thinning of the ozone layer, and the loss of biological diversity. The research in this field deals with how such tendencies could be corrected, how to team up political, economic, technological, and ethical approaches so as to open the way to world security of the environment.[14]

These analyses are in line with the conservationist discourse of the 1980s, which intended to explain how the preservation of planetary resources contributes to the survival of humanity and to development perpetuity.[15] In an era when international relations were modeled after the military (and especially nuclear) strategies of the two superpowers, ecologists also began talking in similar terms, mapping out in this context a world conservation "strategy."[16] As today's concerns are focused more on questions of security than on strategic confrontation, ecologist discourse takes this evolution into account and it now hinges on the notion of security of the environment, thus voicing once more the vocabulary ordinarily used in reference to the state entity. However, the problems to settle and the objectives to attain remain the same, so much so that at the level of content the expressions "conservation strategy" and "security of the environment" are, from an ecologist's perspective, interchangeable in many ways.

B. The Environment as a Component of State Security

Other studies deal with the question differently: the environment is looked upon as an independent variable, whereas state security, which is involved, acts as a dependent variable. Environmental security thus takes on the meaning of "the environmental component of national security."[17]

These studies are based on the premise that because of their source or their gravity, environmental problems can be a hazard to a state's national security. It would actually be affected to various degrees depending on whether such problems trigger social unrest, political instability, economic difficulties, threats to territorial integrity, diplomatic tension, or even open warfare. The studies are therefore interested in the conflict dimension of interstate relations as they crop up in political, economic, diplomatic, or military confrontations stemming from environmental antagonism, be it local, regional, or global in nature.[18]

The correlation between national security and its environmental component is examined from two angles. The first deals with environmental problems as the *main* insecurity factor. The scenarios are based either on confrontations springing from local or regional ecological conflicts (transborder pollution, overexploitation of a common resource, and such) or on a transformation of power relationships within a region—or among several regions—as a result of major environmental disturbances (climatic changes, desertification, ecological accidents, and the like). The second deals with environment problems as an *accessory* insecurity factor. In such cases, environmental antagonisms threaten a state's national security indirectly by helping to exacerbate preexisting political, economic, social, or military tensions or conflicts, or by adding a new dimension to them.[19]

In the eyes of many political scientists, this is a seductive approach because it blends well with the view of international relations that attributes rational behavior to state entities engaged in power politics. On the other hand, as most of the studies up to now have dealt with empirical applications of the environmental security concept without first trying to circumscribe and then clarify its more strictly theoretical dimension, scholars have not yet really tackled within such an approach the thorny question of the adequacy of the existing conceptual tools for suitably handling the problematics of environmental insecurity in a framework originally set up to analyze interstate violence in its most direct form.[20]

III. Redefining the Concept

As we have just seen, the concept of environmental security finds an extremely wide application range. This inevitably raises certain difficulties when the time comes to make the concept operational. To whom and to what does it actually apply? Within which context? Under what conditions? To which ends? Though very real, these operational challenges are nevertheless not insurmountable and can be met by first defining the concept in clear, simple, and precise terms. This has never been done before and it may explain why there is still, in the field, considerable confusion about the meaning of environmental security. From our vantage point, this fundamental task is made easier if the defining efforts are carried out within the broad context outlined by the basic tenets of realism worded at the outset of this text.

According to dictionaries, the word *security* denotes a situation of tranquility resulting from an absence of danger. From a realist perspective, this tranquility situation is valued, in the current international system, in relation to the dominant entity, the state. Indeed, regardless of the school of thought we call upon to analyze and comprehend the international system, it must be conceded that it still largely remains an interstate affair. Consequently, in matters of security, everything gravitates in a comparable proportion around the notion of state security. Be it national, regional, or international, this security is about the ability of states, and the societies within, to maintain their independent identity and their functional integrity.[21] It thus embodies the organization and the political, economic, social, military, and environmental conditions required for the survival of state entities and the international community they constitute. Even though these conditions are inextricably linked, the fact remains, as rightly pointed out by Barry Buzan,[22] that each defines a focal point within the security problematic and a way of ordering priorities. Although each of them cannot be interpreted in isolation from one another, researchers should be careful not to mistake one for another and consider as a threat to the environmental security of a state an event that in reality calls into play another component of its national security. This risk of confusion is particularly great with the military component as the notion of threat is still so very much a part of it. To avoid falling into this trap, it should be emphasized that any threat to environmental security must project a nonconventional character, meaning that it must be positioned beyond the classical range of military confrontation. If today the application range of environmental security appears so extensible it is primarily because researchers were unable to distinguish between the environmental and the military components. Environmental security responds to a dynamic of its own: while the other components of national security may be active, they should not dominate the stakes.

Within this context, it is suggested that the concept of environmental security be redefined in the following terms: *For a state, it represents an absence of nonconventional threats against the environmental substratum essential to the well-being of its population and to the maintenance of its functional integrity.* This definition incorporates four elements whose reach we will now try to circumscribe.

The first element is the state. According to the definition, environmental security must be expressed within a state perspective. This is of course very much in tune with realism, within which the concept of environmental security cannot simply be given a "security of the environment" meaning without the risk of distorting the very notion of security since the latter remains the state's prerogative and should be understood as such under all its aspects be they political, economic, social, military, or environmental. Consequently, the suggested definition distances itself from the trend that uses the expression "environmental security" beyond the framework of state security to designate overall environment safeguarding measures. Moreover, according to current international law, the very idea of a world environment common to all simply does not exist (Taylor 1992). Wanted or not, this law still divides the earth along strictly political—and we would add realist—borders: the territories of states and the zones located beyond national jurisdictions. The concepts of sovereignty and territorial integrity—in which are steeped the notion of national security—are the very foundations of international environmental law. In that system, "there is no direct protection of the environment per se."[23] It is states which have rights and which assume obligations. Hence the necessity, if only on a practical level, to extend the preponderance of these entities within the very core of the environmental security concept. For realists, whose studies essentially focus on the behavior of power-maximizing states, this preponderance is not only necessary but also natural considering that transnational forces, public opinion, ideologies, international organizations, and small or developing states have, by themselves, very little impact on the preservation of the power equilibrium within the existing international system.

Paradoxically, this reference to the state in the definition also makes it possible to envisage environmental security within a more global perspective. Indeed, faced with major environment stakes such as climatic changes, the thinning of the ozone layer, and the loss of biological diversity, which all have the potential to threaten a great many states, it would be possible to measure, through the national securities involved, the impact of these stakes on the environmental security of a region or even on that of the international community as a whole. The suggested definition would therefore offer the possibility of interpreting the concept of environmental security according to one and the same code be it on a local, regional, or international level.

Moreover, notice has to be taken of the fact that the definition leaves it up to the state, as the target of the threat, to make its own judgment as to its insecurity. This is an important point, at least from a realist perspective, because it allows for consideration not only of real changes occurring within the national security of a state, but also of perceptions of changes that could provoke the state into action.

The second element of the definition deals with the notion of threat. It involves a key element because any question of security is necessarily expressed around this notion.[24] Besides, this element has the net advantage of being easy to translate in terms of issues. In a general way, by adapting the previously cited words of Richard Ullman to the theme of our study, it can be considered that a threat to a state's environmental security exists once an action or sequence of events directed in a nonmilitary perspective against its environmental substratum threatens the quality of life of its population and/or the range of policy choices available to the government in its main areas of activity.

Such an all-inclusive formulation can be explained on the one hand by the very close link existing between environment questions and a population's quality of life and, on the other hand, by the much more concrete character of the notion of insecurity, as we have already pointed out. A threat to a state's environmental security can be voluntary or involuntary, direct or indirect, and can spring from internal or external sources. However, it should be noted that, in most cases, it is difficult to be specific about a threat, considering the risks of subjective appraisal, when it comes, for example, to its real weight, its degree of contingency, and its veritable consequences. This is even more true when we consider that the propensity to perceive events through the prism of a security threat varies from one state entity to another. Thus we can perceive in environmental security matters an extremely large range of threats which, however, inevitably, creates certain analytical problems such as how to distinguish between a threat and the intangible factors of life on a planet in constant evolution; understanding why an environment conservation measure initiated by some could be seen by others as a threat to their environmental security; or to dismiss threats which, due to their military nature, call into play another component of a state's national security.

This last point is particularly significant in that the suggested definition deliberately dismisses conventional threats, meaning of a military nature. Consequently, according to the definition, conflicts involving, for example, the appropriation or control of a resource (such as water or oil) cannot be understood as falling into the investigation area of environmental security. Resources are traditional stakes in warfare: the battle for their control and appropriation involves the military component of national security rather than its environmental component, and even more so in that such appropriations cannot be carried out without territorial conquest (thus without "threats" in the conventional sense of the word). Similar reasoning can be applied to the use of environment resources for military purposes. This type of activity consists of artificially triggering destructive natural phenomena such as hurricanes, earthquakes, tidal waves, or fires, or of using resources as environmental weapons (massive spills of crude oil, contamination of waterways, etc.). All these situations presuppose a state of war as well as an armed intervention by the state under attack. Consequently, there would be no question of perceiving or analyzing them other than within the diagnostics of military security. Under the realist paradigm, there is no possible application of green diplomacy here.

The third element involves the environmental substratum meaning the ecosystems, resources, and natural cycles whose degradation, nonsustainable exploitation, or modification could threaten the environmental security of a state and, hence, of a region or the international community as a whole. Here the word *ecosystem* is understood in its etymological sense of "systems maintaining life and its diversity." For example, it denotes the forests (tropical and northern), the coastal zones, the prairies and the savannahs, the cold winter semideserts, the river basins, and humid zones. For any state, preserving these systems represents not only an investment in the maintenance and upgrading of its population's quality of life, but also an important step in protecting its national interest. It therefore cannot do without the constantly evolving possibilities provided by these systems, which prove to be indispensable under numerous aspects for its own sustainable development or, as realist scholars would put it, for its own national prosperity and security.

The term *natural cycles* refers to the various ecological processes characterizing our planet that are essential to the maintenance of the productivity of ecosystems on which depend the survival and development of states and their societies. This particularly includes the climate, the recycling of carbonic gas by the oceans and forests, the annual cycle of the polar freeze-up, and the stratospheric ozone. Some of these natural cycles are already undergoing changes causing concern over their consequences on food production and health.

As to the word *resources*, it has a double meaning within the context of the definition. It primarily denotes natural resources and in particular those which, although said to be renewable, are nonetheless limited and particularly depletable (for example, fishing and food-producing resources). It also denotes natural elements—water, air, and soil—which are essential to the very survival of any human community. Resources therefore constitute fertile ground for many forms of environmental antagonisms. The fourth and last element of the definition deals with a state's social well-being (the welfare of its population) and the maintenance of its functional integrity. Around these two criteria will be determined, according to the definition, the impact of a threat to environmental security: does it attack the quality of life of the inhabitants of the state involved or does it prevent it from functioning normally in one or the other of its main spheres of activity? In other words, to what degree is the national interest of the state jeopardized by the threat? From a realist standpoint, these questions are important not only because they make it possible, as we have seen, to map out a notion of threat but also because they provide a frame of reference, valid both internally and externally, by which to understand the context within which a state measures its degree of vulnerability so as to decide on whether to favor one component of its national security to the detriment of other components, or to establish a certain balance between each of them.

It can furthermore be seen that these two criteria are of a general order in that they remain the same regardless of the source—

political, economic, social, military, or environmental—of the threat to national security. They are not specific to the environmental component.[25] But their reach nevertheless is sufficiently defined to make it possible, when analyzing an environmental security problem, to provide a better evaluation of the intensity of a threat or to distinguish between a threat and what ultimately constitutes only a normal problem in international relations.

Conclusion

The environment topic invaded the universe of international studies wherein two schools of thought dominate. The first revolves around how the spectacular elevation of pollution to the ranks of planetary problems has helped broaden the scope of international studies while developing new perceptions, new concepts, and new approaches. The second revolves around how environmental problems simply portray the traditional political and economic stakes of international relations in a different light.

This difference in viewpoints is nowhere more evident than when assessing the concept of environmental security. For some it involves a new concept that harbors a major potential for discoveries both on a theoretical and practical level. For others it involves a classical stake—security—which has found a new raison d'être through the environment. However that may be, a certain ambiguity exists in one camp as well as in the other when it comes to the empirical reality that the term *environmental security* is supposed to convey.

In view of the importance that environmental problems took on over recent years when it came to formulating new challenges in the area of international security, the purpose of this chapter is to contribute, by drawing on the realist paradigm, some elements that would make it possible to define more precisely the application range of the environmental security concept.

It now remains to be seen where it could lead us. For those who favor the institutionalist approach, the suggested definition raises the question of whether states will actually be able to guarantee their environmental security or whether, instead, they should set up an international institution to handle this near-insurmountable task. As we have seen, the unreliability of existing international mechanisms for managing the environment raises a reasonable doubt about the capacity of the current interstate system to set up sufficiently dynamic and efficient institutions. Even though it has a future, the road leading to the institutionalization of interstate relations is just as filled with pitfalls as we once again learned through the efforts expended to set up the Commission on Sustainable Development. Indeed, the existing institutions have achieved their best results to date in playing the very realist game of power relationships through debt/nature exchanges for example.

As for those favoring the liberal approach, the suggested definition raises the question of whether the pursuit of environmental security is not becoming an unattainable goal per se in view of the fact that the required complex and sustained cooperation could never see the light of day within a system dominated by states primarily concerned with their national security. The argumentation tends on the contrary to demonstrate that, in a realist view, states will seek cooperation whenever it directly serves their self-interest. Despite the absence of a supranational authority, cooperation actually remains possible but on condition, of course, that there be a mutually profitable exchange and that the agreed-upon norms be respected as to how the states are to act. It is true that they act first and foremost in line with their national self-interest, which can be defined, at least partially, in terms of power and security. This is actually why we stated that the impact of pollution problems on the environmental security of a region or of the international community as a whole can only be measured in terms of national securities thus threatened. Consequently, this pursuit of national self-interest by states does not mean that things inevitably will go from bad to worse in environment matters. On the contrary, in the current system this could be the only promising road, as seems to be indicated in a recent study on the determining role the pursuit of national interest played in the setting up of two main (and rare) international environmental protection regimes—namely, the ozone layer protection regime and the regime aimed at reducing transborder atmospheric pollution caused by acid rain.[26] It should be pointed out, however, that when pushed to its logical conclusion this reasoning makes one think that if the pursuit of national interest can foster environmental cooperation, then in the final analysis it would remain focused on the concerns—or the self-interest—of the dominant powers.

Within the perspective we have developed, the classical notions of national interest and security are given a new relevance, which perhaps we had overlooked but which we now believe must be recognized.

──────────────── NOTES ────────────────

1. Richard N. Lebow, "The Long Peace, the End of the Cold War, and the Failure of Realism," *International Organization* vol. 48, no. 2 (1994), pp. 249–77.

2. Sverre Lodgaard, "In Defence of International Peace and Security: New Missions for the United Nations," *UNIDIR Newsletter* vol. 24 (1994), pp. 5–11.

3. Charles-Philippe David, "La crise des études stratégiques" [Strategic Studies in Crisis], *Études internationales* vol. 20, no. 3 (1989), pp. 503–15.

4. Barry Buzan, "Change and Insecurity: A Critique of Strategic Studies," *Change and the Study of International Relations: The Evaded Dimension*, Barry Buzan and R. J. Barry Jones, eds. (New York: St. Martin's Press, 1981); Lester Brown, "An Untraditional View of National Security," *American Defense Policy*, 5th ed., John Reichart and Steven Sturm, eds. (Baltimore, Md.: Johns Hopkins University Press, 1984); Jessica Mathews, "Redefining Security," *Foreign Affairs* vol. 68, no. 2 (1989), pp. 162–77; and Richard Ullman, "Redefining Security," *International Security* vol. 8, no. 1 (1983), pp. 129–53.

5. Norman Myers, "Environment and Security," *Foreign Policy* vol. 74 (1989), pp. 23–41.

6. Richard Ullman, "Redefining Security," op. cit., p. 133.

7. Neville Brown, "Climate, Ecology and International Security," *Survival* vol. 31, no. 6, pp. 519–32; Fen Osler Hampson, "Peace, Security and New Forms of International Governance," *Planet Under Stress: The Challenge of Global Change*, Constance Mungall and Digby J. McLaren, eds. (New York: Oxford University Press, 1990); Patricia M. Mische, "Ecological Security in an Interdependent World," *Breakthrough* (Summer/Autumn 1989), pp. 7–17; Norman Myers, "Environment and Security," *Foreign Policy* vol. 74 (1989), pp. 23–41; Michael Renner, "National Security: The Economic and Environmental Dimensions," *Worldwatch Paper No. 89* (Washington, D.C.: Worldwatch Institute, 1989); and Arthur H. Westing, "The Environmental Component of Comprehensive Security," *Bulletin of Peace Proposals* vol. 20, no. 2 (1989), pp. 129–34.

8. Gerald O. Barney, ed. *The Global 2000 Report to the President of the United States* (New York: Pergamon Press, 1980).

9. Philippe Moreau Defarges, "Environnement et relations internationales" [Environment and International Relations], *Rapport annuel mondial sur le système économique et les stratégies [World Annual Report on the Eco-*

nomic System and Related Strategies], Thierry de Montbrial, ed. (Paris: Éditions Dunod, 1990).

10. Nico Schriver, "International Organization for Environmental Security," *Bulletin of Peace Proposals* vol. 20, no. 2 (1989), pp. 115–22.

11. Sverre Lodgaard, "The Transformation of Europe and the Gulf War: Implications for the Research Agenda," *UNIDIR Newsletter* vol. 4, no. 1 (1991), pp. 7–9; Arthur H. Westing, "The Environmental Component of Comprehensive Security," *Bulletin of Peace Proposals* vol. 20, no. 2 (1989), pp. 129–34; and Arthur H. Westing, "Environmental Security and Its Relation to Ethiopia and Sudan," *Ambio*, vol. 20, no. 5 (1991), pp. 168–71.

12. Sverre Lodgaard, "The Transformation of Europe and the Gulf War: Implications for the Research Agenda," op. cit., p. 64.

13. Arthur H. Westing, "The Environmental Component of Comprehensive Security," op. cit.

14. Sverre Lodgaard, "The Transformation of Europe and the Gulf War: Implications for the Research Agenda," op. cit., p. 64.

15. Daniel B. Botkin, *Discordant Harmonies: A New Ecology for the Twenty-first Century*. (New York: Oxford University Press, 1990); Rice Odell, *Environmental Awakening: The New Revolution to Protect the Earth* (Cambridge, Mass.: Ballinger, 1980); and Jonathan Porritt, *Sauvons la Terre [Saving the Earth]* (Brussels: Casterman, 1991).

16. International Union for the Conservation of Nature and Its Resources (IUCN), *World Conservation Strategy* (Gland: IUCN Press, 1980).

17. Johan J. Holst, "Security and the Environment: A Preliminary Exploration," *Bulletin of Peace Proposals* vol. 20, no. 2 (1989), pp. 123–28; Thomas Homer-Dixon, "On the Threshold: Environmental Changes as Causes of Acute Conflict," *International Security* vol. 16, no. 2 (1991), pp. 76–116; and Philippe Moreau Defarges, "Environnement et relations internationales" [Environment and International Relations], op. cit.

18. Philippe Moreau Defarges, "Environnement et relations internationales" [Environment and International Relations], op. cit., pp. 376.

19. Johan J. Holst, "Security and the Environment: A Preliminary Exploration," op. cit., p. 123.

20. Daniel Deudney, "The Case Against Linking Environmental Degradation and National Security," op. cit.

21. Barry Buzan, "Environment as a Security Issue," *Geopolitical Perspectives on Environmental Security*, Paul Painchaud, ed. (Quebec: Laval University, GERPE Occasional Paper 92–05, 1992), p. 4.

22. Barry Buzan, "Environment as a Security Issue," op. cit., p. 4.

23. Prue Taylor, "The Failure of International Environmental Law," *Our Planet* vol. 4, no. 3 (1992), pp. 14–15.

24. Richard Ullman, "Redefining Security," op. cit., pp. 133–35.

25. Richard Ullman, "Redefining Security," op. cit., p. 133.

26. Detlef Spring and Tapani Vaahtoranta, "International Environmental Policy," *International Organization* vol. 48, no. 1 (1994), pp. 77–105.

5

The Case for DOD Involvement in Environmental Security

Kent Hughes Butts

I am persuaded that there is also a new and different threat to our national security emerging—the destruction of our environment. The defense establishment has a clear stake in countering this growing threat. I believe that one of our key national security objectives must be to reverse the accelerating pace of environmental destruction around the globe. —Senator Sam Nunn

With the end of the Cold War, national security analysts began reexamining the definition of national security. As a result of this analysis, the notion that security necessarily

meant the threat of violence and military conflict was widely abandoned and the definition of a national security interest was broadened to include other threats to U.S. interests, such as economics and
the environment, and in 1991 the environment was included in the
National Security Strategy of the United States (NSS) for the first
time. In that document, the president pointed out that the failure to
manage the earth's natural resources in ways that protect the potential for growth produces stress that is already contributing to political conflict.[1] Subsequent NSSs address the importance of environmental change, linking it decisively to sustainable economic growth
and regional stability and naming it a direct threat to U.S. national
interests.[2]

The inclusion of the environment in the NSS demonstrates a
common national and international consensus of the environmental
dimensions of conflict, growth and development, health, and political
stability. Environmental issues such as clean air, desertification, and
natural resource access have a cross-border component that has contributed to international tension and conflict, and therefore, threatens U.S. national interests.

By usefully conceiving national security to include the environment, the NSS suggests that organizations with traditional national
security roles, such as the Department of Defense (DOD), should expand their supporting strategies to include environmental objectives. Indeed, the 1997 NSS, reflecting multiagency environmental
initiatives and major environmental speeches by former Secretary of
State Christopher, former Secretary of Defense Perry, and former Director of Central Intelligence Deutch, defines a strategy of "shaping
the international environment" in which the diplomatic, military,
and economic elements of power preemptively act to prevent or reduce threats to U.S. interests.[3] The U.S. military is well suited to respond to this new national security strategy.

The Clinton administration moved quickly to reinforce the concept of environmental security and the important role DOD would
play in executing the NSS by creating the position of the Deputy
Under Secretary for Environmental Security, and appointing
Sherri Wasserman Goodman to fill it. In her May 13, 1993 statement before Congress, Ms. Goodman defined DOD's environmental
security mission as "ensuring responsible environmental performance in defense operations and assisting to deter or mitigate impacts of adverse environmental actions leading to international instability."[4] She thus recognized DOD's importance to two major

aspects of environmental security, the threat to human health and the contribution to regional conflict. Given the growing importance of environmental security and the administration's vision that DOD will play a major role, it is important to understand what DOD is doing in the environmental security arena, and what further contributions it could make.

The U.S. military has taken great strides to address environmental security issues. Internally, DOD has spent billions of dollars to rectify past environmental problems such as toxic and hazardous waste spills and soil erosion, and to promote biodiversity and conservation. However, in addition to improving its domestic environmental stewardship, the military has accepted the larger mission of "assisting to deter or mitigate impacts of adverse environmental actions leading to international instability." In executing this international environmental security mission, the U.S. military has sent forces to assist in the cleanup of East European military bases, help manage the demilitarization of Soviet nuclear weapons, and established a biodiversity and conservation program in Africa to help nations better manage their fisheries and unique natural wildlife resources. The military organizations of the developing states are relatively well organized and present in all areas of a given country, and have the transportation resources necessary to have a meaningful impact upon issues that threaten their country's environmental security. Thus, DOD has employed the military forces of developing nations to mitigate environmental problems that threaten security with a tangential benefit of maintaining good communication between the United States and the militaries of these countries. DOD need not undertake such activities unilaterally. Other federal agencies, as well as NATO, have the mandate to undertake similar missions.

Environmental security issues threaten the existence of the earth's ecosystem and regional stability. Consensus on the importance of these issues cuts across North-South and East-West lines. Yet taking action to address agreements developed at such environmental conferences as Rio and Cairo inevitably proves difficult because of the inability to monitor, coordinate, and enforce agreement provisions. Because of its vast experience in solving complex environmental problems domestically and in developing and industrialized countries, unique technical and operational capabilities, and global network, the U.S. military can integrate and harmonize efforts to address these critical issues and elevate the overall performance of those actors responsible for their resolution.

The Military's Toxic Legacy

Environmentalists are right to be skeptical about the concept of involving the military in efforts to improve the environment. When Saddam Hussein released millions of gallons of oil into the Persian Gulf he demonstrated the extreme environmental consequences of warfare, and the Cold War left a legacy of environmental disaster that is only now being revealed. The Soviet Union's development programs for weapons of mass destruction were conducted with little regard for human health or natural resources, spreading contamination and chronic health problems from Altay to the Yakutia area of Vilyui. Moreover, plutonium extraction waste was discharged directly into lakes and rivers, and nuclear submarines and ships were sunk at sea and continue to leak fuel and pollute fisheries.[5] While the Soviet's unprincipled behavior is egregious, the United States also has nuclear-weapons-related environmental problems. Fifty billion dollars has already been allocated to clean up nuclear waste at Hanford, Washington. And the Department of Energy, which is responsible for nuclear weapons production, could spend over $200 billion dollars to remediate its weapon production sites and waste.[6]

In the nonnuclear environmental area, DOD has many problems. Because DOD produces and maintains weapons and equipment, it is a major industrial operator and produces industrial waste, much of it toxic and hazardous. By current estimates there are approximately 19,694 hazardous waste sites on 1,722 DOD installations. Some of these are minor, but 93 DOD installations have Superfund sites (the National Priority List of major hazardous waste sites).[7] As a result, DOD has many environmental problems that will cost billions of dollars to correct. To be fair, many of DOD's problems were created in the years before the United States became environmentally aware and enacted comprehensive environmental laws. Nonetheless, the problems exist and reflect badly on the military.

The publicity surrounding the clean-up of its toxic waste sites has caused DOD to reexamine its environmental commitment and programs and has generated positive benefits. DOD discovered that it is cheaper to prevent pollution than to clean it up. It has reduced its hazardous waste pollution by 50 percent and wants to help other countries do the same.[8] Further, it is investing hundreds of millions of dollars into the development of new clean-up technologies that will benefit private and international toxic and hazardous waste cleanup efforts. DOD is also investing $2 billion per year to comply with existing environmental rules and another $2 billion on remediation. DOD has become an instrument with which the U.S. Government can improve environmental security, overseas and domestically.[9] The government and DOD now realize that the environment is a national security issue, and a DOD responsibility.

DOD's Foreign Environmental Role

> We must manage the earth's natural resources in ways that protect the potential for growth and opportunity for present and future generations.... Global environmental concerns ... respect no international boundaries. The stress from these environmental challenges is already contributing to political conflict.
> —1991 National Security Strategy

The National Security Strategy correctly points out the major role that environmental problems play in conflict initiation, particularly in the developing world, which has the highest population growth rates. These growth rates, some approaching 4 percent, can double a country's population in as few as twenty years, depleting resources, threatening the legitimacy of newly democratic governments, and severely testing their ability to satisfy the needs of their people. It also serves as strong pressure for governments to seek solutions at the expense of their neighbors.

For many countries, environmental issues are *survival interests* and therefore potential sources of regional conflict. In the Middle East, for example, the lack of water in Iraq and Syria has the potential to further destabilize the region because Turkish water resource policies threaten the economic vitality and traditional way of life of its downstream neighbors. On the margins of the Sahara and in the Horn of Africa, refugee populations of several million have crossed borders to escape the starvation caused by overgrazing and drought. These refugees may increase the population of the host country by 10 percent or more and promote poverty. They strain the economic and social infrastructure, limit the ability of the government to satisfy demands, and promote disharmony among the indigenous population when refugee organizations disproportionately address the needs of the refugees. Environment security issues are eroding political stability and are hence appropriate issues for the military to address.

The end of the Cold War and the need to sustain the conditions necessary for peace and to eradicate poverty and environmental problems, have changed the focus of the joint State Department/DOD Security Assistance Program for the developing world. Encouraged by Congress, the United States has moved from selling heavy military weapons and equipment to supporting nation building, environmental sustainability, and small-scale unit training, with the hope that this will help overcome such barriers to democracy as ethnicity.

Because developing countries frequently have several national groups within their political borders, the potential for ethnic conflict exists. When the government is composed disproportionately of a particular ethnic group, then other ethnic groups often assume that the resources of the country are not being evenly distributed. When the country experiences economic difficulties, the government will likely be accused of unfairly reducing benefits to the regions of the country populated by the other ethnic groups, with the potential to undermine political stability. To the degree that the developed world can provide economic and health assistance and aid in managing the natural resources and environment of the country, it creates a more positive milieu in which these groups may seek national identity and a feeling of belonging, helps the government satisfy systemic demands, and reduces the potential for conflict.

DOD is creatively using Security Assistance Programs in some developing countries to encourage host military forces in the nontraditional roles of actively promoting biodiversity, natural resource conservation, and environmental management. Quite often in the developing world, the domestic military's role in governmental policymaking is much greater than in the developed world. The military is frequently a better-organized, better-trained and more technologically sophisticated element of the government than are other comparable organizations. Moreover, the military is generally present in all regions of the country, including the frontier, where a feeling of national identity may be absent. Because ethnicity is such an important factor in developing world politics, the military quite often will have a more regionally representative and ethnically diverse population than does the highest level of government. The military may thus promote a sense of national identity by performing environmental and sustainable economic development missions.

When the developing world military can be encouraged to foster sustainable development, it promotes governmental legitimacy and political stability and serves as a useful and viable environmental resource. This fact is recognized in the new DOD environmental security mission of mitigating or deterring environmental problems that could lead to political instability.

Recognizing that poverty is the chief cause of political instability, the U.S. military has been assisting the developing countries' militaries to promote sustainable development and to maintain their natural resource base through components of the Security Assistance Program. The U.S., African Biodiversity and Conservation and Civic Action Programs, for example, provide funding, guidance, and engineering review to the host government military for nonmilitary projects designed to conserve natural resources and benefit the civilian population, often in remote areas of the country. Moreover, they create in the local military a resource with which the host government can address pressing environmental problems.

In the critical area of natural resource conservation, the U.S. program is helping the African littoral states control the problem of international fish poaching. For decades, large fishing fleets from European and other nations have come within the economic zones and twelve-mile limits of African countries and have literally raped the ocean floor, destroying the habitat while vacuuming large schools of fish from the sea. In Namibia, for example, a five-month incursion of Spanish trawlers fishing for hake cost the country $100 million and resulted in long-term damage to the fisheries.[10] This poaching debilitates the habitat, threatens the carrying capacity of the fishing areas and impairs the livelihood of indigenous fishermen and the foreign-exchange earning potential of governments. By providing the means for these countries to patrol their own economic zones (through the use of patrol boats, small observation planes, radios, and training in interdiction and international law procedures), the U.S. military has helped curtail poaching in African littoral waters, restoring the pride of the local governments and promoting regional cooperation on a critical environmental issue. This solution addresses a particularly virulent international environmental problem because the endangered fish stocks migrate up and down the coast across international boundaries.

In Fiscal Year (FY) 1991, for example, Congress made available $15 million under the Foreign Assistance Act to help the militaries of African countries protect and maintain wildlife habitats and to institute sound wildlife management, fishery, and conservation programs. This DOD-managed program concentrates on wildlife habitat maintenance by constructing roads in game parks, developing and building bridges and dams, and reinforcing antipoaching efforts for game parks and fisheries. It also develops a host country capacity to protect marine and terrestrial wildlife and fisheries. In Botswana, for example, $2.4 million has been allocated to train antipoaching units and to purchase air boats for use in the famous Okavango Swamp, as well as two small airplanes for aerial surveillance.[11]

Cape Verde is receiving $1.7 million to purchase a fifty-foot patrol boat that will lead antipoaching efforts within its economic zone. Equatorial Guinea, Burundi, and the Central African Republic will also receive jeeps, communications equipment, and training to develop viable antipoaching units for their game parks. In Madagascar, a quarter of a million dollars is being provided for the army to restore a major game park in that politically troubled country. In Malawi, $1.5 million has been allocated to further improve its well-established wildlife management program, which has developed elephant herds large enough to require culling.[12]

Because of its success, in FY 1993 Congress allocated an additional $15 million for the Biodiversity and Conservation Program. These attempts to support biodiversity and natural resource preservation in the developing world are examples of how military-to-military contacts and the resources of the Department of Defense can prove useful in the effort to improve the world's environment. It also helps DOD, because the Congress is increasingly critical of U.S. support to developing world militaries, fearing that such aid might promote military dictatorships. As a result, Security Assistance Programs are being heavily scrutinized and reduced.

By realigning the Security Assistance and Military-to-Military Programs with the new National Security Strategy, administration and congressional intent, and stressing environmental security, DOD is reducing the likelihood of regional conflict and maintaining important relationships. DOD has responded to congressional willingness to fund environmentally related Security Assistance Programs, using them to enhance economic sustainability and the political stability of struggling governments. This contributes to regional political stability in the developing world. It also maintains the military-to-military contacts that provide base access, overflight, staging, and logistical support agreements, the strategic objectives that enable DOD to project power during conflicts in the increasingly unstable developing world. However, DOD need not approach the problem on a unilateral basis.

In 1991, the North Atlantic Council redrafted the Alliance Strategic Concept. This redrafting recognized that threats to allied security were less likely to occur from the aggression of traditional East Bloc enemies, but were more likely to occur from economic, social, environmental, and political problems; and that these problems could "lead to crisis inimical to European stability and even to armed conflicts."[13]

The Alliance decided that the objectives of its strategic concept could best be achieved through political means such as the use of dialogue and cooperation for the purpose of reducing the risks of conflict, fostering confidence-building measures, and maintaining military-to-military contacts with Eastern and Central Europe and the former Soviet Union. This new Strategic Concept provides the framework in which lessons learned from the U.S. environmental security program can be translated to the NATO mission and used to address the serious environmental problems that pose a direct threat to European security.

Extreme environmental degradation is a direct threat to European security because it jeopardizes political stability in regions that are essential to Europe, which obtains much of its petroleum and strategic minerals from the former Soviet Union, the Middle East, and Africa. All three of these regions face difficult environmental problems for which too few resources exist to effect resolution. As a result, Europe runs the risk of losing access to these natural resources, faces growing waves of refugees fleeing those decimated areas for the physical and economic security of the European continent, and must be concerned with a traditional military threat from the former East Bloc countries. Still in possession of nuclear and other weapons of mass destruction and their waste materials, the governments of these countries are fragile and unable to satisfy the demands for internal environmental security placed upon them by their newly formed democratic constituents.

The environmental degradation that affects these regions can be managed, but not without the technical and managerial expertise of the developed world. The Middle East is at risk from overpopulation and conflict over scarce fresh-water resources. The former Soviet Union and East and Central Europe are beset with toxic and hazardous waste spills, widespread air pollution, and the threat from the continued poor management and storage of weapons of mass destruction and their waste products. Africa is also threatened by overpopulation and the resulting erosion of topsoil and desertification, which have sent millions of refugees across political borders.

To address environmental problems that threaten European interests, NATO could build upon the work done by its Committee on the Challenges of Modern Society (CCMS) and the framework of the North Atlantic Cooperation Council (NACC) to establish an environmental security assistance program that would address the environmental problems of the regions important to European security.[14] This organization would be headed by a general officer and consist of teams that would interact with specific countries and regions, tailoring their assistance to provide the expertise necessary to mitigate the specific problems of that area.

The NATO teams could draw upon the technical, training, and managerial expertise of NATO and the DOD programs to address such important issues as hazardous waste assessment and mitigation design, environmental threat monitoring, water resource management, the teaching of natural resource conservation practices, disaster relief planning and training, the restoration of military facilities, and the management of weapons of mass destruction disposal. The NATO environmental security assistance program could also combine with other allied environmental security assistance programs such as the work being done by the United States European Command (EUCOM), and could also serve as a clearinghouse for environmental proposals to be funded by outside donors. This clearinghouse function could establish priorities, coordinate the currently unfocused efforts of multiple donor agencies and governments to address environmental problems in these critical areas, and concentrate environmental resources against the threats that are most relevant to European security. The program's objectives should be tied to the strategic aims of the European community and the alliance, and should also reflect the goals of the multilateral lending institution programs from which much of the environmental mitigation monies come.

This program would benefit both parties. The receiving countries, which typically lack the technical, managerial, and administrative expertise to execute these missions, will have their militaries trained to become environmental security resources and will have difficult environmental problems properly addressed. NATO benefits from reducing the sources of potential instability, while maintaining the military-to-military contact and communication that can help to diffuse misunderstandings and conflict. Moreover, NATO would further demonstrate its relevance to the modern security environment and take advantage of the requests of the former Soviet Union and East and Central Europe cooperation partners to increase their affiliation with NATO.

It makes sense to use military resources to address the root causes of potential conflict early and inexpensively rather than waiting until conflicts such as Somalia become full blown and then undertake missions that are costly both in financial and human resources. The commitment of environmental security assistance teams can help mitigate the causes of political instability and future conflict and preclude DOD and NATO out-of-area combat missions.

DOD's Domestic Environmental Role

DOD is first and foremost a resource that can be used to address U.S. national security objectives. As the definition of these objectives has broadened to include such environmental issues as human health, resource management, and global ecology, DOD has been found to be a unique asset with which to achieve these objectives. Many of the environmental skills DOD has to offer the international community were developed in its domestic program, which exemplifies why the military should be a valuable resource to a state's environmental security strategy.

DOD has a substantial domestic environmental mission which, in the current environmentally sensitive milieu, it must execute almost flawlessly if it is to retain control of its training areas and installations as well as the support of the American people. DOD spends billions of dollars on its domestic environmental mission, frequently in cooperative programs with other federal agencies such as EPA, the National Oceanic and Atmospheric Administration, and DOE, as well as state fish and game departments, Native American Tribal organizations, universities, and environmental organizations. None of these organizations can match the resources DOD can bring to bear on environmental problems, or unilaterally achieve the same success. Many of the nation's environmental problems are so large in scope or cost that they border on the unmanageable. Because of its size and resources, DOD has the capabilities to address many of these problems.

Few realize the size and power of the Department of Defense. It employs three million people with an annual budget of $250 billion. Domestically, DOD manages twenty-five million acres and over one thousand significant installations, in addition to the thousands of National Guard and Reserve Component properties.[15] These installations compose the equivalent of national parks, recreation areas, vast old-stand forests, and desert areas. Most have complex, urban developments of residential, industrial, and commercial areas that require municipal management practices. Unlike municipalities, these installations cannot lobby and receive exemptions for environmental noncompliance. They must comply or face fines, and the funds for environmental compliance largely come from the existing DOD budget.

The size of DOD's land holdings makes it one of the largest environmental managers in the United States. This management role has increased over the years because the extensive development occurring on private land often leaves the only remaining natural habitats on DOD installations. Thus, DOD lands have become de facto game preserves for endangered species. Natural disasters and private development in the South, for example, have destroyed the habitat of the Red-Cockaded Woodpecker and have made the old-stand timber on Forts Bragg, Polk, and Benning, and Camp LeJeune particularly important to the survival of this endangered species. Installation commanders must now conduct realistic, combat simulation training while managing flourishing herds of game, ensuring the nesting process of endangered species is undisturbed, and enhancing wetlands. The responsibility for environmental stewardship will continue.

Although military land is used for training, by law it must be rotated, maintained, and its wildlife habitats protected. Training areas, such as Fort Carson, Colorado, and Hohenfels, Germany, that were once abused and needlessly eroded, have been reseeded and are carefully managed with the assistance of the local civilian wildlife administrators. This is good for DOD trainers because they train in a realistic, living environment, not a dust bowl. Thus, while national security and economic reasons will prevent a major reduction in the volume of military lands, this is not necessarily a negative environmental development. DOD lands provide a buffer against development and must now be managed for multiple use in strict accordance with state and federal regulations and the oversight of natural resource agencies. Doing so conserves the physical environment where it may otherwise be lost. Because the Department of Defense manages an area roughly the size of Tennessee, it has the capacity to evenly apply national natural resource laws over a substantial area of the United States.

Perhaps no other area of environmental concern more clearly demonstrates the value to the United States of having the Department of Defense execute national environmental standards than natural resources and endangered species protection. Because national level environmental or natural resource programs have a regional focus, they present unique coordination challenges across agency, state, and municipality lines, and difficulties with financing and resource allocation. The Coastal America program, for example, is designed to manage and protect coastal resources. Participants include the Environmental Protection Agency, U.S. Fish and Wildlife Service, U.S. Geological Survey, Minerals Management Service, the National Oceanic and Atmospheric Administration, the National Park Service, and the U.S. Army Corps of Engineers. The program's goals include reversing habitat loss and degradation, reducing pollution from nonpoint sources, and remediating contaminated sediments.[16] With such diverse goals, many state and local agencies as well as private organizations and the public must actively participate, and often, individual locales find themselves unable to develop the resources necessary to perform the requirements for which they are responsible. For this reason it is important to have a strong federal presence to demonstrate that the program can work. DOD is playing such a role.

The Chesapeake Bay is one of the regions selected by Coastal America in which the Department of Defense is heavily involved. The largest estuary in the United States, the Chesapeake is threatened by the population of the Bay basin, which will double its 1950 level by the year 2020.[17] Pollution from industry, agriculture, and urban areas had threatened to destroy the Bay's productive capacity and in 1975 it was singled out for rehabilitation by the Environmental Protection Agency (EPA). DOD was the first federal agency to sign an agreement with the EPA to participate in restoring the Bay.

DOD's participation is significant and demonstrates why it is a major factor in national environmental security. Its presence ensures that some 350,000 acres of the Chesapeake Bay watershed and sixty-five major DOD installations will be sharing common goals and standards of performance. As a federal agency, DOD can allocate resources that most municipalities cannot. In FY 1990, for example, DOD spent approximately $50 million on erosion control, natural resource management, pollution prevention, and waste reduction in order to improve the water quality of the Bay.[18]

DOD is providing further assistance to this program through the capacity of the U.S. Army Corps of Engineers to share environmental expertise and assistance with local, municipalities and other federal, local, and state agencies that are involved in designing remediation efforts in their own areas of responsibility. The Corps of Engineers Waterways Experimentation Station developed, and is sharing with the other Coastal America program partners, the Chesapeake Bay computer model used to prioritize and manage the resolution of common Bay problems.

Particularly important to the natural resource aspect of the Chesapeake Bay Initiative is the fact that the Army alone manages approximately 225,000 acres of Bay watershed, 84 percent of which is undeveloped.[19] The DOD installations are thus serving as natural resource preserves where the habitat for indigenous waterfowl and other fauna and flora remain free from development pressures. What other areas can one find in populous regions (where development is both ongoing and deemed in the best interests of the local economy) where such acreage can be assured of preservation?

This is not the only program or agreement concerning natural resources into which the Department of Defense has entered. An agreement between DOD and the Nature Conservancy ensures that biological diversity on DOD installations is documented and maintained and that endangered and threatened species are identified. This is particularly important because DOD installations contain over 250 threatened or endangered species. Rather than being on the defensive with environmental groups, DOD is now reaching out to these sources of expertise in the quest for technical advice and assistance in developing procedures for planning and conducting natural resource improvement programs.

When environmental concerns in the United States began to focus on the health and environmental effects of toxic and hazardous waste products, DOD, like many corporations, found that it had substantial areas that required remediation. DOD's formal efforts to come to grips with its toxic and hazardous waste problem were initiated in 1984 with the Defense Appropriations Act. Under this act, the program known as the Defense Environmental Restoration Program (DERP) was established and funded through a new transfer account known as the Defense Environmental Restoration Account (DERA). The DERP has two major programs: the Installation Restoration Program (IRP) and the Other Hazardous Waste Operations Program (OHW). Under the IRP, contamination at DOD installations and formerly used defense properties are investigated and, when necessary, the clean-up process is begun. Under the OHW program, research and development and demonstration programs are initiated that reduce the rates at which DOD hazardous waste is generated.

The funding that DOD has allocated to the DERA program has increased constantly since 1984 and reflects both the resources required by DOD to effect this clean-up and its commitment. In 1984, the first year of the DERA program, $150 million was allocated for environmental restoration. By 1991 DOD was spending over $1 billion annually to clean up contaminated sites. For FY 1994 the figure jumped to $1.9 billion, and for FY 1996, DOD allocated $1.2 billion.[20] The billions of dollars that DOD has invested in its clean-up program has created a vast pool of technical and managerial expertise that can be made available to resolve domestic and international toxic and hazardous waste problems.

While DOD once hid behind the construct of Sovereign Immunity when dealing with state and local enforcement agencies, it has now developed a model, the Defense and State Memorandum of Agreement, that facilitates the open and active participation of states in cleaning up DOD installations.[21] DOD has approached all states and asked that they participate in this memorandum of agreement program. More than forty states and/or territories have favorably responded. Under this program DOD pays the states to participate in and monitor the clean-up processes at DOD installations. In FY 1991 for example, approximately $16 million was given to state regulatory agencies to facilitate their assistance in evaluating and providing oversight for installation restoration program actions. DOD is also spending substantial amounts of money to clean up former bases designated for closure (BRAC). In FY 1997, $724 million is being invested in the BRAC programs for former defense bases.[22]

The wisest use for environmental funding is to prevent pollution in the first place. DOD set a goal of reducing hazardous waste production by 50 percent between 1987 and 1992. By 1993, DOD had achieved its goal. In FY 1994, DOD invested $275 million in hazardous waste minimization projects.[23]

The resources that DOD has dedicated to the daily compliance with environmental laws have risen substantially over the years. In FY 1991 DOD spent in excess of $800 million to conform to federal, state, and local environmental regulations and laws. In FY 1992 it increased its compliance funding to some $1.9 billion and in FY 1997 the figure rose to $2.02 billion.[24]

A primary benefit of having DOD actively involved in environmental work is that it brings national level resources to bear on state and local environmental problems. DOD brings a breadth of experience and the ability to transfer regional solutions to other parts of the United States. DOD offers a combination of local expertise and management and high-powered national level financial and technical resources to the environmental arena, spending approximately $5 billion per year on environmental improvement.

No other organizations or associations of organizations have the regional presence, management expertise, or resources to execute these environmental missions with the success of DOD. Using DOD resources for local or regional environmental improvement is a nascent phenomenon. As more associations of state and local governments turn to DOD for help, DOD's role and contributions will expand.

Conclusion

In spite of substantial inertia and a reluctance in some academic circles to give up the Cold War era focus of security studies on deterring military conflict, senior national security policymakers at DOD, state, and the CIA have recognized environmental variables as threats to U.S. national security interests, from democratization to regional stability. This has been made clear in National Security Strategies and numerous Presidential Decision Directives. As an element of national power traditionally used to address threats to national security, the military is an appropriate resource to involve against environmental threats. It is time to look beyond narrow perceptions and historical fears concerning the role of the military, and creatively use this element of power to further U.S. interests in the current, dynamic milieu.

Local and state government, the administration, and Congress increasingly view the Department of Defense as a positive agent of environmental security. DOD is perhaps the best resourced of all federal agencies performing an environmental mission. The military is well organized and can bring its organizational abilities to bear on local environmental problems. It is present in every state, in all regions around the globe, and it has substantial financial, human, research and development, and technological resources with which to contribute to the nation's environmental security objectives. Moreover, properly managed, DOD environmental R&D efforts will create products that could be marketed overseas by U.S. corporations, improving international security while enabling U.S. firms to expand their market share in the lucrative environmental market. DOD has established environmental cooperation agreements with conservation groups and has made significant contributions to federal, state, and local environmental programs. In addition, DOD's international programs are encouraging the armed forces of developing countries in nontraditional roles that include biodiversity and environmental improvement, health care, and sustainable economic development. DOD has substantial potential for further positive contributions to environmental security without compromising operational readiness. It is a resource that can support and magnify the accomplishments of organizations responsible for global and domestic environmental improvement.

————————— NOTES —————————

1. U.S. Government, *National Security Strategy of the United States* (Washington, D.C.: U.S. Government Printing Office, 1991), p. 22.

2. U.S. Government, *National Security Strategy of the United States* (Washington, D.C.: U.S. Government Printing Office, 1997), p. 5.

3. Ibid, pp. 6–9.

4. Sherri Wasserman Goodman, *Statement before the Subcommittee on Installations and Facilities, of the House Committee on Armed Services*, May 13, 1993.

5. "Geological Exploration Directorate Maps Moscow Radiation Hazards," *Novaya Yezhednevnay, JPRS-TEN-93-023*, September 15, 1993, pp. 28–30.

6. Keith Schneider, "New Mission at Energy Department: Bomb Makers Turned to Cleanup," *The New York Times*, August 17, 1990.

7. Department of Defense, *Defense Environmental Restoration Program (DERP): Annual Report to Congress for FY 1994* (Washington, D.C.: U.S. Government Printing Office, 1994), p. 2.

8. Gary Vest, *Interview*, Washington, D.C., May, 1997.

9. Department of Defense, *Defense Environmental Restoration Program*, op. cit., pp. 3–6.

10. Helmoed-Romer Heitman, "Future Structure of Navy Envisaged," *FBIS-AFR-92-206*, October 23, 1992, p. 6.

11. Scott Fisher, "Biodiversity Country Projects," Office of the Assistant Secretary of Defense, International Security Affairs, Africa Region, 1992, p. 1.

12. Theresa Whelan, *Interview*, Washington, D.C., September, 1994.

13. NATO, "The Alliance's New Strategic Concept," *Press Communique S-1*, NATO Press Service, November 7, 1991, p. 3.

14. Wendy Greider, "NATO Works with Cooperation Partners on Environmental Problems," *NATO Review* (April, 1993).

15. William H. Parker III, "Environment Moves to Front Burner," *Defense 90* (March/April 1990), pp. 21–32.

16. Coastal America Partnership, *Coastal America Partnership for Action*, Washington, D.C., 1991, p. 2.

17. Army Environmental Office, *Department of the Army Chesapeake Bay Year-End Report*, 1990.

18. H. Leonard Richeson, "Keynote Address," Hazardous Materials Control Research Institute, 7th National RCRA/SUPERFUND Conference, St. Louis, Mo., May 2, 1990.

19. Donald L. Robey and Larry J. Laurer, "A Time-Varying Three-Dimensional Hydrodynamic and Water Quality Model of Chesapeake Bay," *The Military Engineer*, August, 1991, p. 50.

20. Department of Defense, *Defense Environmental Restoration Program: Annual Report to Congress for FY 96* (Washington, D.C.: U.S. Government Printing Office, 1996), p. 4.

21. Department of Defense, *Defense Environmental Restoration Program*, 1991, op. cit.

22. Department of Defense, *DOD Environmental Security Programs (Program Summary)*, Washington, D.C., February 21,

23. Department of Defense, *Defense Environmental Restoration Program*, 1994, op. cit., p. 9.

24. Department of Defense, *DOD Environmental Security Programs (Program Summary)*, Washington, D.C., February 21, 1997.

6

The Case for Comprehensive Security

Eric K. Stern

This chapter makes the case for a comprehensive conceptualization of security, including a prioritized environmental component. Broadening concepts of security remains controversial and the issue of how to treat the interface between the realms of security and the environment has generated heated debate among practitioners and scholarly observers of international relations. At the root of this controversy is a fundamental question: How should scholars, citizens, and practitioners negotiate the conceptual terrain between these domains?

The comprehensive security approach offers an appealing alternative to both environmentally negligent military-centered approaches and more radical globalist approaches. Comprehensive security reflects and corresponds to the multiple responsibilities of governance delegated by individuals to collective institutions at all levels of social aggregation, from the local, state or provincial, national, regional, and to the global. In dealing with these multiple responsibilities, decision-makers at all levels are continually faced with difficult value tradeoffs and complex interdependencies among issues. These tradeoffs and interdependencies may be probed vigorously or ignored and avoided.[1]

Unfortunately, there are strong psychological and organizational tendencies to deny and avoid rather than rigorously probe value conflicts. As a result, concern for the protection of the environment has traditionally played a minimal or even nonexistent role in the selection of military-security postures, with devastating consequences for the environment and the health of many citizens.[2]

Citizens, scholars, and practitioners alike are best served by the comprehensive security approach because it accentuates and reveals, rather than compartmentalizes and disguises, the interrelationships and potential tensions between different values and realms of human endeavor. This approach promises to facilitate the resolution of value conflicts by encouraging the search for integrative solutions that may transcend apparent value conflicts. Where such solutions do not appear attainable after vigorous examination, this approach may at least encourage the sober and measured acceptance of value conflict and, indirectly, provide a basis for more responsible policymaking.[3]

This chapter proceeds in three steps. First, recent proposals to extend the concept of security to encompass environmental and other types of issues are examined. Second, three lines of criticism—orthodox realist, antistatist globalist, and statist reformist—of linking environment and security are examined and refuted. Third, the case is made that a comprehensive security approach can help produce more balanced, integrated, and environmentally sound security policy-making.

Broadening the Concept of Security

For several decades a heated debate has been underway among practitioners and scholars of international relations in the subfields of security studies and peace studies. This discussion has resulted in the generation of a cornucopia of alternative security concepts. An incomplete list of modifiers that have been proposed includes the following:

collective, cooperative, mutual, international, regional, world, global, economic, societal, democratic, environmental, ecological, and comprehensive. A detailed critical review of these proposals is beyond the scope of this chapter, but it is possible to examine some of the key points of convergence and contention.[4]

Despite this diversity of perspectives, there is a conceptual baseline uniting the diverse efforts to broaden concepts of security: in a nutshell, the security policy problem is involved with coping with potential *threats* to and the establishment of conditions conducive to the promotion of core *values* cherished by individuals or communities. As Barry Buzan has suggested, focusing on "core values" is useful because analysts are often ambiguous regarding the stability, subjectivity/objectivity, and internal consistency of such values.[5]

While falling short of consensual definition of security, a focus on the interplay between threats and values provides a common denominator that cuts across much of the diversity of formulations and conceptualizations identifiable in the literature.[6] However, it is important to note that it is the high level of generality and the inclusion of both negative (threat to values) and positive (promotion of values) variants of security that builds bridges to many of the existing conceptualizations. At the same time, as we shall see, this baseline glosses over significant disagreement over which aspects should be emphasized and the specific scope and content that should be assigned to these general terms.[7]

Varieties of Expanded Security Concepts

With these general thoughts in mind, let us examine several of the more prominent and well-developed proposals to expand the concept of security. The least controversial is the proposal to include economic phenomena as part of security.[8] More sweeping proposals aim at incorporating environmental, political, and societal (culture and identity) dimensions into the concept of security.[9]

In one of the first attempts to formulate an expansive security concept, Richard Ullman moved considerably beyond the traditional military-centric conceptualization. Ullman defined a "threat to national security" as "an action or a sequence of events that (1) threatens drastically and over a relatively brief period of time to degrade the quality of life for the inhabitants of a state, or (2) threatens significantly to narrow the range of policy choices available to a state or to private, nongovernmental entities (persons, groups, corporations) within the state."[10] Ullman also specifically identified "a drastic deterioration of environmental quality caused by sources from either within or outside a territorial state" as one such threat.[11] One problem with this definition is that its second part is overly expansive if taken literally. One may legitimately wonder whether a significantly narrowed range of choice for corporations or individuals is sufficient to constitute a threat to *national* security. A personal or corporate bankruptcy might well suffice to significantly narrow the range of choice of the affected individual or organization, but such a situation constitutes a threat to individual or corporate security rather than national security. Despite this difficulty, Ullman's argument was a seminal contribution to the security debate.[12]

Perhaps the most fertile area of recent innovation in security concepts has been environmental. Building on early work by Lester Brown and others,[13] numerous analysts in the late 1980s advocated conceptualizing environmental problems as security threats. For example, Jessica Tuchman Mathews, described the process of successive reformulations of the definition of U.S. security as part of ebb and flow of the international context and the U.S. position in the world.[14] She notes that "in the 1970s the concept was expanded to include international economics as it became clear that the U.S. economy was no longer the independent force it had once been, but was powerfully affected by economic policies in dozens of other countries," and observes that "global developments now suggest the need for another analogous, broadening definition of national security to include resource, environmental and demographic issues."[15] In the 1990s the end of the Cold War, the rise of public concern over environmental issues, and the advocacy efforts of environmental security analysts have given this concept of broadened security widespread visibility.

In analyzing environmental threats to security, Bengt Sundelius has usefully distinguished between threats that arise from potentially malignant actors, and those that result from structural processes which create relations of vulnerability or instability.[16] Others argue that changes in technology and social organization have created new kinds of dangers, such as climate change and nuclear proliferation, that threaten human well-being on a global scale.[19] Environmental threats to security are often, though not always, structural in character. Ozone depletion and climate change are examples of longer-term threats to security derived from anthropogenic and natural processes rather than the potentially malicious intentions and capabilities of other human actors.

Ozone depletion is widely thought to constitute a clear and present danger to the health of human beings and animals, as well as to agriculture.[18] The evidence suggests that health problems such as skin cancers and eye disorders are likely to result from heightened levels of ultraviolet radiation due to ozone depletion. Recent research findings suggest that a hole in the ozone layer over the Arctic region, similar to the one previously identified over the Antarctic region, may be developing.

While the threat posed by climate change is less well understood, the ecological and social disruptions that may result from this phenomena are thought to have extremely negative implications for security, particularly for individuals living in less developed regions of the globe. For example, the Alliance of Small Island States (AOSIS) has actively sought to influence negotiations over a global warming convention, arguing that rising ocean levels thought to be a result of the greenhouse effect pose a grave threat to the national security of these nations. Even relatively modest rises in ocean levels threaten to inundate these islands with sea water, rendering them uninhabitable. AOSIS members have called for drastic measures on the part of the industrialized world to reduce emissions of greenhouse gases.[19]

Security, the State, and Common Security

These characterizations of problems as security threats pose several key questions apart from the issue of "what kinds of threats." Other contested issues are the appropriate object (or demand side) of security and the appropriate level of social aggregation at which security is to be sought (the supply side). Analysts also disagree about what kinds of strategies should be adopted in seeking security.

In conventional usage of the phrase "security policy," a preceding adjective is often left implicit. This suggests that a deeper consideration of these conventional assumptions is required. Buzan's influential study of the security concept poses two of the key questions: "what is the referent object for security? And what are the necessary conditions for security?" Buzan observes that "security as a concept clearly requires a referent object, for without an answer to the question 'the security of what?', the idea makes no sense."[20] The reflexive response to Buzan's first question is to privilege the nation or the state, as in the common usage of "national security."

However, making the state the sole or primary object of security is highly problematic, as Buzan points out, because the state itself often poses a serious threat to the well-being of its own individual citizens or ethnic groups. For example, the Soviet Union's infamous KGB (Committee for State Security) was dedicated to the protection of the state against domestic and international foes, and was a major threat to individual citizens through its suppression of civil liberties and Orwellian tactics. Even in Western societies there exists a trade-off between state security and individual liberty. As Ullman observes, "individuals and groups seek security against the state, just as they ask the state to protect them from harm from other states." As a result, "human rights and state security are thus intimately linked."[21] Peter Katzenstein's recent study of the response of security institutions in West Germany to transnational terrorism during the 1970s and 1980s reveals these difficult tradeoffs in a contemporary Western setting.[22] This dilemma of collective identity and institutionalization operates across cultures and ideological divides, and at virtually all levels of social aggregation. For example, gang membership in an inner city neighborhood provides a modicum of protection to gang members, but members become subject to violent sanction if they break the unwritten rules of membership. They are also subject to being drawn into conflict with other gangs and law enforcement authorities.

The international dimensions of security have also been reconceptualized in important ways. Because of the presence of other states in the anarchical international system, the security of individuals living in states is unavoidably entangled in interstate networks of security relationships. In such interstate systems "security dilemmas" can result from uncoordinated unilateral attempts at achieving national security through arms buildups that are counterproductive to security.[23] In order to respond to these problems, the concept of international or "common security" has gained widespread currency. While still largely state-centric, common security emphasizes the mutual interest in avoiding destructive conflict shared by many states.[24]

Recent efforts to broaden and reconceptualize security have also focused on the "supply side"—the issue of determining the appropriate level of social aggregation for efforts aimed at coping with serious threats.[25] While many analysts tend to focus on *strategies* geared toward a particular level, many security threats seem best managed by initiatives at a variety of levels of aggregation, depending on the possibilities and constraints provided by the political context in question.[26] Attempts to mitigate threats to individual human rights deriving from state repression may include organization of political parties and movements, constitutional reform at the state level, regional institution-building, international nongovernmental organizations such as Amnesty International, and global efforts in fora such as the United Nations. Perceived threats of violent attack by the armed forces of other states may be met by preparations for nonviolent resistance, unilateral military buildup, bilateral arms control efforts, or multilateral arrangements such as the nuclear nonproliferation treaty. Attempts to develop strategies to respond to global health threats such as the AIDS virus may be addressed at all levels: voluntary modification of individual behavior, national legislation (e.g., mandatory screening of blood supplies), or through international efforts on the part of the World Health Organization. Clearly, appropriate strategic mixes may vary from issue to issue and from context to context.

The Political Presence of Expanded Security Concepts

Concepts of expanded security have increasingly appeared outside the analyst community in important political settings. Several of these proposals for expanded concepts of security have developed in parallel with, and in some cases in dialogue with, political proposals developed by several international commissions convened under United Nations auspices. In 1980, the Brandt Commission (The Independent Commission on Development Issues) called for a broadening of the scope of the security concept: "Our survival depends not only on the military balance but on global cooperation to ensure a sustainable biological environment and sustainable prosperity based on equitably shared resources."[27] In 1987, the Bruntland Commission (The World Commission on International Development) couched its findings in the language of expanded security: "Environmental threats to security are now beginning to emerge on a global scale. The most worrisome of these stem from the possible consequences of global warming caused by the atmospheric build-up of carbon dioxide and other gases."[28] In 1995, the Commission on Global Governance, chaired by Ingvar Carlsson and Shridath Ramphal, endorsed a refocusing of security to emphasize the environment and individual human rights: "Global security must be broadened from it traditional focus on the security of states to include the security of people and the planet."[29]

Proposals for environment-security linkages are not confined exclusively to the realms of academia and the United Nations, but have increasingly found their way into the vocabularies of senior officials in a wide range of countries, including the United States and Russia.[30] In a passage typical of this new political language, President Bill Clinton employed the vocabulary of comprehensive security:

> Not all security risks are military in nature. Transnational phenomena such as terrorism, narcotics trafficking, environmental degradation, rapid population growth and refugee flows also have security implications for both present and long term American policy. . . . [A]n emerging class of transnational environmental issues are increasingly affecting international stability and consequently will present new challenges to U.S. strategy.[31]

These statements demonstrate that expanded conceptualizations of security have moved beyond the proposal stage to become a substantial and growing part of the discourse of practitioners in contemporary world politics.

Three Criticisms Considered

The attempt to expand conceptualizations of security to include environmental problems as threats has been criticized from a diverse set of perspectives. Three criticisms are of particular note. First are objections raised by what might be termed orthodox realism, the dominant school in American security studies and international relations theory. The orthodox realist view asserts that a relatively restrictive definition of security should be maintained in order to protect the conceptual and substantive integrity of security studies as a discipline. Second, a more radical globalist critique reflects a distaste for the normative and ontological implications of security discourse and argues that it is inappropriate to "securitize" nonmilitary social issues and projects.[32] Third, the reformist view suggests that raising environmental issues to the level of "high politics" is likely to be counterproductive and that the role of environmental factors in generating social conflict is best dealt with under the general framework of conflict studies.

Orthodox Realism

Most mainstream contemporary American realists reject treating environmental problems as security issues and maintain that a narrow definition of national or international security delimited to concern with threats originating in the military capabilities of other states.[33] Realism is a heterogeneous tradition of international theory and practice, and as demonstrated in the earlier chapters in this volume by Deudney and Frédérick, not all varieties of realism are hostile to linking environment and security. Capturing the view held by mainstream American neorealism, Stephen Walt has recently defined the scope of security studies as "the study of the threat, use, and control of military force."[34] He argues that security studies "explores the conditions that make the use of force more likely, the ways the use of force affects individuals, states, and societies, and the specific policies states adopt in order to prepare for, prevent, or engage in war."[35] Walt argues that incorporating nonmilitary phenomena entails an excessive expansion of security studies: "Defining the field in this way would destroy its intellectual coherence and make it more difficult to devise solutions to any of these important problems."[36] Walt's realist orientation is an orthodoxy seeking to protect a core of security studies focused on state and military security issues.

Conceptual purism does not, however, prevent Walt from placing "economics and security" in a prominent position in his proposed "research agenda for security studies."[37] He legitimizes this broadening by referring to linkages between military spending and economic performance, posture resources, and the impact of the military-industrial complex. But Walt's exception to his general posture of exclusion seems less than airtight: once the door is opened to nonmilitary economic activities, it becomes difficult to justify excluding other nonmilitary issues from the security studies agenda.

The environment is no more remote from the traditional miliary core of the field than Walt's approved economic topic. There are a range of research topics on a range of topics located at the intersection of environment and military security.[38] Four links between environmental and military issues are particularly important: environmental contamination from weapons development/production, pollution and influence attempts, the environmental consequences of warfare, and environmental degradation as a cause of conflict. By briefly examining these topics, it will be clear that Walt's posture of exclusion is unjustifiable.

First, the industrial processes associated with the development and production of weapons often have serious adverse effects on the environment. For example, the negative environmental impacts of nuclear testing became widely known already during the 1960s and resulted in the Partial Test Ban treaty which aimed at minimizing nuclear pollution ("fall out") by prohibiting nuclear tests in the atmosphere and the oceans. Concern that even underground testing may pose serious risk stimulated efforts in the 1990s to achieve a comprehensive ban on nuclear testing. The storms of protest directed at France and China for disregarding the informal moratorium on nuclear testing followed by the other nuclear powers is indicative of the magnitude of concern caused by this issue. Severe radioactive pollution has resulted from the operation of facilities for the development and production of nuclear weapons in both the United States and the former Soviet Union.[39] The potential environmental impacts of biological, chemical, and even conventional weapons development, production, and storage also clearly fail to respect a clean divide between environment and security.

Second, deliberate pollution is a significant military and political tool used to influence the behavior of other actors in world politics.[40] Both states and nonstate actors such as terrorist groups may threaten or actually intentionally produce environmental catastrophes in pursuit of political aims or as a means of venting their rage at perceived injustice in the world system. For example, starvation is not only an unintentional side-effect of violent conflict, but a widely used tool of coercion.[41] The Iraqi behavior during the Gulf Crisis and War of 1990–1991 is another important example of deliberate pollution intended to have a military impact. The Iraqi occupiers of Kuwait intentionally induced a significant regional environmental catastrophe. According to T. W. Hawley, the oil spills and oil fires of the Gulf war were "an environmental catastrophe on the order of the explosion at the Chernobyl plant or the lethal release of methyl isocyanate at Union Carbide's chemical plant in Bhopal, India."[42] The threat and intentional inducement of environmental catastrophes bears a striking family resemblance to the threat and use of economic sanctions as a diplomatic instrument, the importance of which Walt concedes.[43]

The intentional inducement of environmental disasters as part of warfare is not a new phenomenon.[44] Deliberate "scorched earth" tactics have been part of the military repertoire since classical antiquity at the very least.[45] Such tactics have been employed in more recent conflicts as diverse as the American Civil War and World War II in order to deny territory or the economic potential of the land to an adversary.[46] The massive application of toxic defoliants (such as Agent Orange) to jungle areas by American forces during the Vietnam war adversely affected the environment, not as an unintentional side-effect, but as a primary aim. Recent conflicts in Afghanistan,

Cambodia, and Bosnia have left landscapes strewn with millions of landmines and booby traps. This insidious and lingering environmental hazard continues to exact a tremendous toll in human suffering and economic costs associated with mine clearing and care for countless victims.[47] Military activity also often results in various forms of unintentional pollution. Such pollution ranges from vehicle exhausts to radioactive contamination resulting from the detonation of armor-piercing shells tipped with depleted uranium.

Another potential interface between these issue areas is the contested proposition that environmental degradation and resource competition contributes to the outbreak of war and other forms of violence. Many analysts have argued that the economic and social dislocations associated with climate change may result in civil and international wars.[48] Homer-Dixon and his colleagues have attempted to rigorously specify the paths to conflict, suggesting that simple scarcity conflicts, group-identity conflicts, and relative-deprivation conflicts may increase the probability of resort to inter- and intrastate violence.[49] Lodgaard's proposal for a regime of environmental confidence and security-building measures (CSBMs) to mitigate the tendencies to violence identified by Homer-Dixon further points to the artificiality of a rigid separation between the environment and traditional security concerns.[50] These phenomena clearly meet Walt's narrow criteria of "conditions which make the use of force more likely."[51]

Still another interface concerns opportunity costs, multiple purpose capabilities, and conversion. Opportunity costs are incurred when assets invested in military-security projects become unavailable for other potential uses, a phenomenon Buzan sees as part of the "defense dilemma."[52] Such tradeoffs include the traditional "guns for butter" tradeoffs between defense and private consumption, as well as between defense and education, health care, environmental clean-up, and the development of environmentally benign technologies. A second aspect of this interface is the multiple purposes that can be served by military capabilities.[53] For example, military units are commonly employed for rescue- and disaster-relief missions which may involve responding to natural cataclysms (such as earthquakes, floods, volcanoes, hurricanes, etc.) or to technological catastrophes with serious environmental consequences (such as nuclear accidents or oil spills). The military is well suited for such missions because of its transportation, engineering, and medical capabilities, logistical readiness, and organizational discipline. Finally, conversion refers to the possibility of reallocating existing military assets and capabilities to other uses, which can entail either the transfer of capabilities out-side the military sector or the allocation of responsibility for new en-vironmental clean-up missions to existing military organizations.

To sum up, the attempt on the part of orthodox realist scholars to isolate security studies from environmental phenomena is ulti-mately unconvincing. The environmental security agenda falls com-fortably within the parameters of the security studies agenda, even when that agenda is defined in traditionally military-centric terms.

Globalist Criticisms

A second line of criticism, voiced by Daniel Deudney and others, also opposes linking environmental concerns with the concept of national security, but for very different reasons. While the realist orthodoxy wants to privilege the state and nation, Deudney's globalist perspec-tive sees environmental politics as subversive of the state, and the state as subversive of the emergent global environmental political sensibility. In a frequently cited article, Deudney makes three basic claims. First, he claims that it is "analytically misleading" to charac-terize environmental degradation as a threat to national security, be-cause "the traditional focus of national security has little in common with either environmental problems or solutions." Second, he argues that "the effort to harness the emotive power of nationalism to help mobilize environmental awareness and action may prove counter-productive by undermining globalist political sensibility." Taking the position that the security notion is still primarily linked to the states system and "national" security, Deudney argues that mixing environ-ment and security betrays the ethos of the environmental movement: "Environmental degradation is not a threat to national security. Rather, environmentalism is a threat to 'national security' mindsets and institutions." He further claims that dressing the environmental program in the "blood-soaked garments of the war system" will undermine environmental "core values" and create "confusion about the real task at hand." Third, he doubts that environmental degrada-tion will be a significant cause of interstate wars.[54] These are power-ful and sophisticated arguments shared by many environmental-ists,[55] but they fail to do justice to the actual thrust of the concept of environmental security.

It is beyond the scope of this chapter to seriously challenge Deudney's claim that environmental degradation is not likely to lead to interstate war. In this author's view, the evidence is not yet clear on this point and serious observers may disagree on their estimate of this probability. However, it is certain that war and preparations for war lead to environmental degradation on a very large scale. Thus, the traditional focus of national security actually has much in common with environmental problems and possible solutions.

The most developed environmental security analysis, provided by Thomas Homer-Dixon and his colleagues, indicates that environmental degradation is more likely to exacerbate social conflict within states rather than among them.[56] Does such conflict belong on the security agenda? Deudney's globalist orientation actually undermines the foundations of the domestic-international distinction. Recent history, most notably the tragic civil wars in Somalia and Rwanda, and the conflicts associated with the breakup of the former Yugoslavia, indicate that this distinction is becoming less significant to the international community. Extremes of civil violence or other episodes of large-scale human suffering, brought into the living room by a mass media with instantaneous communications of global reach, have evoked calls for intervention by major powers and regional and international organizations such as NATO and the U.N. Therefore, serious intrastate conflicts linked to environmental degradation are increasingly on the security agenda of major actors in world politics.

How serious is Deudney's concern that incorporating environmental issues into the security concept may undermine the globalist political sensibility characteristic of the environmental movement? A key question here is how closely the concept of security is bound to the state. Buzan's conceptualization of security, which challenges the privileged status of national or state security, provides a more realistic approach. Buzan argues that security is not sought or found at a single level of social aggregation, but rather must refer to the individual, national, international levels. He observes that "the concept of security binds together individuals, states, and the international system so closely that demands to be treated in a holistic perspective," and that "a full understanding of each can only be gained if it is related to the other two." Given this, "attempts to treat security on any single level invite serious distortions of perspective."[57]

Security strategies and arrangements entail balancing enduring tensions between these several levels at once. As a result, an increasing sensitivity to environmental threats by national communities and a resulting incorporation of environmental values into national security policy does not necessarily undermine more global efforts to protect the planetary environment. On the contrary, it seems likely that efforts at these two levels may well complement each other. The most likely scenario for the next half century is one of a mixed system of governance characterized by an uneasy distribution of authority between autonomous subnational entities, national governments, and emergent supranational institutions, such as regional and global formal organizations, regimes, and a developing body of international law.

Some scholars have begun to use the term *new medievalism* to understand the character of the post–Cold War order. This concept is particularly appropriate in capturing the politics of Europe, with its uneven processes of integration and disintegration.[58] This mixed system will continue to experience tension between national sovereignty and the transnational or global character of many environmental problems (of both natural and anthropogenic origin) for the foreseeable future.

Furthermore, Deudney's characterization of common security as a marginal phenomenon and an unstable conceptual foundation for environmental security is also questionable.[59] The CSCE process, which culminated in the institutionalization of significant confidence-building and security measures in the European region, and more recently in the establishment of an Organization for Security and Cooperation in Europe (OSCE), indicates that this concept is less marginal and more robust than Deudney suggests. In the years since the publication of Deudney's article, supranational security concepts have clearly become more broadly grounded in the world of practitioners and scholars alike.

To sum up, Deudney's critique, by equating security exclusively with the "national-security-from-violence" problematic, tends to discount the significance of a broad range of parallels and interrelationships between environmental values and military security. In addition, his approach prematurely dismisses the viability and persuasiveness of conceptual innovations building upon the notion of security and acknowledging the interdependence that actually characterizes the problem of "security from violence" to nearly the same extent as the problem of protecting the global environment.

Reformist Criticism of Environmental Security

Another relatively distinct line of criticism, which might be labeled "reformist" has recently been formulated by Marc Levy.[60] Levy makes a commendable attempt to specify the central terms involved[61] and to assess soberly the costs and benefits associated with integrating environment and security. Three arguments warrant examination. First, Levy holds that "the assertion that many environmental problems constitute security risks is correct, and is of very little importance."[62] He argues that two key threats—ozone depletion and climate change—do appear to constitute "environmental problems that currently pose a direct physical harm to U.S. interests," but he questions whether anything is gained by labeling them security issues.[63] Second, Levy argues that taking these environmental issues out of the realm of "low politics" where significant progress has been made, may actually be counterproductive. Third, Levy recommends that the study of indirect threats produced by environmentally induced conflict be left to researchers studying more general processes of social conflict, instability, and war. These researchers should incorporate environmental factors into their more general models of conflict escalation and deescalation. Overall, Levy essentially defends continued compartmentalization and advocates more research within the traditionally defined subfields of international relations and comparative environmental policy.

Although Levy makes a number of good points, his argument suffers from several serious limitations. First, Levy, like Deudney, deliberately focuses exclusively on the issue of national security and for all practical purposes does not even consider the implications of linking environmental concerns with supranational security concepts. Furthermore, Levy seems to conflate more general types of conceptual issues with the specific configuration of U.S. interests. For example, his discussion of the degree of seriousness of the climate-change threat is based on the magnitude of possible deleterious changes and on U.S. resources and potential coping capabilities. However, if one examines the question of climate change as a threat to small island states, which have more limited resources and which face the risk of total inundation from even modest rises of sea level, there are obvious links with the fundamental national security goal of survival.

Also problematic is Levy's claim that it is counterproductive to raise environmental issues to the status of "high politics." Levy challenges the assumption that assigning an issue to the realm of high politics automatically increases the likelihood that forceful action will be taken. But his argument rests primarily upon a single historical example—policy with regard to protecting the ozone layer. Levy claims that progress was made on this issue by "sub-cabinet officials operating out of the limelight and with little Congressional meddling."[64] He claims that "it is hard to escape the conclusion that we probably saved more lives by treating the stratosphere as low politics than as high politics." Yet, is it really possible to draw far-reaching conclusions on the basis of a single case of alleged policy success?

It may be the case that other issues, arising under other circumstances, might well be more effectively and rapidly addressed by a "high-politics" route. Issues requiring significant investment and research, and public sacrifice, might require higher-level and more public leadership. For example, it required the national mobilization initiated by President John F. Kennedy's challenge to the nation to launch the Apollo moon program. In an observation which seems to contradict the main thrust of his position, Levy argues that climate change, in contrast to the relatively straightforward ozone problem, is "much more like the problem of containing the Soviet Union; it requires a grand strategy to guide actions in the face of distant, uncertain threats, and an overarching commitment from high levels of leadership to stay the course through the ebbs and flows of popular sentiment."[65] Thus, for this important environmental issue, Levy calls for a high-politics type solution. Third, and perhaps most seriously, Levy delimits his concern to only "links from processes of environmental degradation to deterioration in security positions," and excludes from consideration "connections that run in the opposite direction (from use of force to deterioration of environmental quality)."[66] Nor are links between preparations for the use of force and environmental degradation considered. Thus, his overall conclusion is based upon an examination of only part of the case made by advocates of environmental security.

The Advantages of Comprehensive Security

A pragmatic approach is needed to help policymakers avoid the Scylla of environmental radicalism and the Charybdis of environmental denial. As we have seen, some analysts have attempted to keep environmental affairs off the security agenda. They have reached these conclusions for a variety of different reasons, none of which seem compelling. Others have proposed placing concern for the environment in a privileged position, in effect toppling the threat of international violence from its traditionally hegemonic position at the core of the security discourse and replacing it with a new environmental "king of the mountain."[67] Neither of these solutions seem

appropriate and neither promise to contribute to dealing constructively with the problem of competing values and interests at stake in
security policymaking.

There is a widespread realization that a unifying conceptual
platform is needed from which to overview the complex relationships and tradeoffs among domains of life and values held by political communities such as nation-states. Rather than privileging
military threats, environmental threats, or any other kind of
threats a priori, we should attempt to "level the playing field" in
order to better appreciate the side-effects associated with particular choices and how best to allocate scare collective resources. Unfortunately, serious consideration of the psychological and organizational constraints on policymaking suggests that achieving such
a unifying overview may be difficult, making suitable concepts even
more valuable.

Value Conflicts in Policy

In a highly acclaimed work addressing the academic-practitioner
nexus, Alexander George identified the perennial problem of value
complexity in policymaking. Policy problems "are laden with competing values and interests" and "the standard textbook model of rationality cannot be employed in such instances, because the multiple values embedded in the policy problem cannot be reduced to a single
utility function that can then be used as criterion for choosing among
options."[68] Even identifying the values embedded in a particular
issue is a demanding process that requires probing facilitated by critical dialogue. It is not uncommon for highly salient values to remain
hidden throughout a decision process only to become embarrassingly
obvious with hindsight. Even where a range of competing values are
identified by a decisionmaking process, there are strong organizational and cognitive pressures toward suppressing value conflict
rather than confronting it. For example, many organizational theorists contend that organizations tend to manage value conflict by ignoring it.[69] Similar tendencies have been identified at the individual
psychological level.[70]

Several major psychological theories and empirical research
findings suggest that value conflict is stressful for most individuals
because they have a subconscious need for cognitive consistency.[71]
Yugo Vertzberger has argued that "people see the social environment
as consistent and balanced." As a result "they believe that a policy

that serves one value is also likely to contribute to other values, thus avoiding consideration of value tradeoffs," and "even salient internal contradictions are often ignored."[72]

In order to deal with these problems, Ralph Kenney has proposed replacing conventional "alternative-focused thinking" with a radically different decision paradigm: "value-focused thinking."[73] In this decision paradigm, "significant effort is devoted to articulating values" prior to other activities. Then "the articulated values are explicitly used to identify decision opportunities and to create alternatives." Kenney argues that such a value-focused orientation will help "create better decision situations with better alternatives, which should lead to better consequences."[74] Such an approach helps to deal constructively and proactively with value tradeoffs. Another important step, and one where academics can make an important contribution, is providing policymakers, journalists, and the public with conceptual tools that encourage examination of tradeoffs and interdependencies operating in the interface between issue areas.

The concept of comprehensive security formulated by Barry Buzan, Arthur Westing, and others promises to serve this important integrating function by explicitly recognizing the multiple and interrelated dimensions and core-value clusters of domestic and international life. Buzan's formulation is particularly useful because it includes military, economic, environmental, societal, and political dimensions. Strategies to promote values associated with these domains and to cope with perceived threats may take a variety of forms and range from self-help to institutionalized cooperation. In this approach, both military and environmental affairs are granted the same basic status as dimensions of a broader comprehensive security concept.

This approach also helps frame a research agenda for further investigation. As demonstrated earlier, there are significant parallels and interactions at the interface of environmental and security domains, and these should be aggressively explored. While the relationships between environmental degradation and social conflict within and across national boundaries remains uncertain, it is certain that peacetime and wartime military activity are major causes of environmental degradation. Intentional environmental destruction as a means of warfare or political struggle is a significant and recurring tragedy that straddles the divide between these social domains as conventionally defined. Issues of opportunity cost, multipurpose capabilities, and conversion also deserve rigorous investigation.

Comprehensive Security and the Nation-State

A comprehensive approach to security can also help to appropriately situate issues concerning the relationship between the nation-state and the environment. It is not useful to regard the security concept as inextricably linked to the nation-state because problems of security and security policymaking cut across the ladder of social aggregation. While the notion of security forged in the international relations discourse has customarily been linked to the nation-state, its gravitational force is not insurmountable. The trend toward disassociating security and the nation-state, already prevalent in world politics, appears strong enough to avoid the dangers of falling down the slippery slope toward a parochial statism that Deudney, Dalby, and others fear.

A comprehensive approach to security can also help integrate the key fact that attempts to enhance security for individuals entails coping with serious tensions among multiple levels of social aggregation and identity. Blind allegiance to the nation-state system and inability to contemplate shifts of sovereignty "upward" or "downward" needlessly limit the range of institutionalized or ad hoc security strategies, which may be brought to bear in averting or coping with military and nonmilitary threats to individual and collective well-being. The steps that have been taken by many West European states to enhance their security through regional integration in the European Union is a powerful example of pragmatic policy that has transcended the traditional conceptual boundaries of international theory.

Despite its limitations, the nation-state remains a central actor on the world stage. This poses important risks but also presents opportunities. On the one hand, there are no guarantees that attaching environmental values to national security will not exacerbate environmental security dilemmas resulting from myopic conceptions of the national interest, as Deudney and Dalby fear.[75] On the other hand, emphasizing environmental values at a variety of levels including the national may help drive home the insight that self-help and nationally based strategies will probably not prove a sufficient basis for coping with the environmental threats now on the horizon. Short of the overthrow of the states system by imperial conquest or popular resistance (neither of which seems imminent), reallocation of sovereignty away from the nation-state require decisions by governments as part of an evolving international practice.

The incorporation of environmental considerations into a broader view of national security policy can support rather than hinder attempts to deal with emerging environmental problems on a more global level.

Conclusions

A comprehensive security concept is needed because of the existence of subtle and not so subtle tradeoffs between values emphasized in various issue realms. Citizens, scholars, and practitioners alike benefit from a conceptual infrastructure that highlights such tradeoffs rather than hiding them. Compartmentalizing issues and maintaining simplistic high-low politics distinctions contributes to unbalanced policymaking. Externalities resulting from choices taken in one issue area are likely to have unrecognized consequences for values associated with other issue areas. Just as it has become increasingly accepted that environmental impacts should be considered as part of economic decisionmaking, the potential environmental costs associated with particular military-security postures should be explicitly factored into decisionmaker calculations.[76]

This has rarely been the case in the past and should not be taken for granted in the future. The concept of comprehensive security promises to provide citizens and practitioners with a framework conducive to managing these tradeoffs in a more responsible fashion. This is far more than rhetoric. Placing the military and environmental dimensions of security within a broader context of comprehensive security will require some semantic adjustment. However, the benefits of such adjustment promise to outweigh the costs.

John Dewey has suggested that our previous understandings and conceptualizations provide a basis for comprehending the novel features of contemporary life: "We cannot lay hold of the new, we cannot even keep it before our minds, much less understand it, save by the use of ideas and knowledge we already possess. But just because the new is new it is not a mere repetition of something already had and mastered. The old takes on new color and meaning in being employed to grasp and interpret the new."[77] As Dewey points out, we have no choice but to make use of the old in our attempts to understand the new. In our attempts to better understand the new kinds of threats facing us in the modern age, it is only natural to use familiar concepts previously deployed in a different fashion as a point of departure. This will inevitably alter our understanding of these concepts. Still, all things considered, this change promises to enhance our understanding of security rather than spoil it.

———————————————— **NOTES** ————————————————

Earlier versions of this chapter were presented at the Conference on Environmental Change and Security (University of British Columbia, 1993), the Nordic International Studies Association Inaugural Meeting (Oslo, 1993), and the Swedish Institute for Interntaional Affairs (1994). I would like to thank participants in those sessions for useful comments, especially Walter Carlsnaes, Daniel Deudney, Magnus Ekengren, Jan Hallenberg, Thomas Homer-Dixon, Magnus Jerneck, Richard Matthew, Ulrika Morth, Gunner Sjostedt, Bengt Sundelius, Lisa Van Well, and Jacob Westberg. Thanks also to Andreas Behnke for taking the time to write a sharp critical essay, which helped identify parts of the argument needing clarification, and to the anonymous reviewers of this volume for their suggestions.

1. Lindblom develops the concept of "social probing" to describe attempts by laymen, scholars, and practitioners alike to explore and develop an understanding of the obscure and confusing social world around them Charles Lindblom, *Inquiry and Change*, (New Haven, Conn.: Yale University Press, 1990), pp. 29–44.

2. For a useful bibliography on the topic, see the Environmental Change and Security Project Report, "Environment and Security Debates: An Introduction," Issue 1 (Washington, D.C.: Wilson Center, 1995), pp 102–105.

3. The concepts of value-conflict resolution, avoidance, and acceptance are drawn from Alexander L. George, *Presidential Decisionmaking in Foreign Policy: The Effective Use of Information and Advice* (Boulder, Colo.: Westview Press, 1980), pp. 28–34.

4. For example, see Barry Buzan, *Peoples, States and Fear: The National Security Problem in International Relations* (Chapel Hill: University of North Carolina Press, 1983); Barry Buzan, *People, States and Fear: An Agenda for International Security Studies in the Post–Cold War Era* (Boulder, Colo.: Rienner, 1991); Neta C Crawford, "Once and Future Security Studies," *Security Studies*, vol. 1, no. 1 (1991), pp. 283–316; and Simon Dalby, "Security, Modernity, Ecology: The Dilemmas of the Post–Cold War Security Discourse," *Alternatives* vol. 17 (1992), pp. 95–134. For an overview of the different approaches, see Joseph Nye and Sean Lynn-Jones, "International Security Studies: A Report of a Conference on the State of the Field," *International Security* vol. 12, no. 4 (1988), pp. 5–27.

5. Core values are subjectively defined and politically contested or contestable. As a result, they are at least potentially changeable. Furthermore, severe contradictions are likely to emerge across multiple central values. Tradeoffs and conflicts among such values are to be expected. Yet, this is not to say that high degrees of intersubjective consensus regarding ends or means cannot emerge at particular junctures and endure for prolonged periods of time. Core values such as survival, autonomy, prosperity, and health are commonly widely held and perennially at or near the top of the political agenda. This does not, however, prevent the emergence of conflicts over which value or values to prioritize at any given point in time. Nor does it preclude substantial conflict over which strategies to employ in promoting one or more of these values. For further discussion, see Barry Buzan, *People, States and Fear: An Agenda for International Security Studies in the Post–Cold War Era* (Boulder, Colo.: Rienner, 1991).

6. For a sampling of the myriad attempts to define the slippery concept of security, see Barry Buzan, *People, States and Fear: An Agenda for International Security Studies in the Post–Cold War Era* (Boulder, Colo.: Rienner, 1991), pp. 16–17.

7. Those accepting one part of the baseline might object violently to another. It should also be noted that a limitation of this baseline is that it tends to conceal the linguistic-political implications of alternative conceptualizations of security. For example, see Murray Edelman, *Constructing the Political Spectacle* (Chicago: University of Chicago Press, 1988).

8. Nye and Lynn-Jones, "International Security Studies," p. 5. Swedish researcher Nils Andren may be identified as an often overlooked pioneer in this respect, voicing his call for "total security policy" emphasizing linkages between military and economic contingencies a decade and a half before the latest round of the security debate got underway during the early 1980s. Nils Andren, "In Search of Security," *Cooperation and Conflict* vol. 4 (1968), pp. 217–39.

9. These include N. Brown, "Climate, Ecology, and International Security," *Survival* vol. 31, no. 6 (1989), pp. 519–32; Richard Ullman, "Redefining Security," *International Security* vol. 8, no. 1 (1983), pp. 129–53; H. Haftendorn, "The Security Puzzle: Theory Building and Discipline Building in International Security," *International Studies Quarterly* vol. 35, no. 1 (1991), pp. 3–17; Arthur Westing, *Comprehensive Security for the Baltic: An Environmental Approach* (London: Sage Publications, 1989); Ian Rowlands, "The Security Challenges of Global Environmental Change," in Brad Roberts, ed., *U.S. Foreign Policy after the Cold War* (Cambridge, Mass.: MIT Press, 1992), pp. 207–22; Barry Buzan, M. Kelstrup, P. Lemaitre, and Ole Waever, *Identity,*

Migration, and the New Security Agenda in Europe (London: Pinter, 1993); and Michael Klare and Daniel Thomas, *World Security: Challenges for a New Century*, 2d ed. (New York: St. Martin's Press, 1994). For a complete overview, see Keith Krause and Michael C. Williams, "Broadening the Agenda of Security Studies: Politics and Methods," *Mershon International Studies Review* vol. 40 (1996), pp. 229–54.

10. Ullman, "Redefining Security," p. 133.

11. Ullman, "Redefining Security," p. 134.

12. For example, Marc Levy has recently attempted to build upon Ullman's conceptual foundation with this definition: "A threat to national security is a situation in which some of the nation's most important values are drastically degraded by external action." Marc A. Levy, "Is the Environment a National Security Issue?" *International Security* vol. 20, no. 2 (1995), p. 40.

13. Lester Brown, "Redefining National Security," Worldwatch Paper number 14, 1977.

14. Jessica Tuchman Mathews, "Redefining Security," *Foreign Affairs*, vol. 68 (1989); and "The Environment and International Security" in Klare & Thomas, eds., *World Security: Challenges for a New Century* (New York: St. Martin's Press, 1994), p. 274.

15. Jessica Tuchman Mathews, "The Environment and International Security," p. 274.

16. Bengt Sundelius, "Coping with Structural Security Threats," in Höll, ed, *Small States in Europe and Dependence* (Vienna: Braumüller, 1983), pp. 281–305; Bengt Sundelius, ed., *The Neutral Democracies and the New Cold War* (Boulder, Colo.: Westview Press, 1987), especially pp. 23–27; and Ullman, "Redefining Security," p. 134.

17. For an overview of proliferation, see Zachery Davis and Benjamin Frankel, eds, "The Proliferation Puzzle: Why Nuclear Weapons Spread (and What Results)," Special issue of *Security Studies* vol. 2, nos. 3 and 4 (1993).

18. Lynton Caldwell, *International Environmental Policy: Emergence and Dimensions*, 2d ed. (Durham, N.C.: Duke University Press, 1990), pp. 262–63.

19. Paul Lewis, "Island Nations Fear a Rise in the Sea," *New York Times*, February 2, 1992, p. A3; and Nicholas D. Kristof, "In Pacific, Growing Fear of Paradise Engulfed," *New York Times*, March 2, 1997, p. A1.

20. Buzan, *People, States and Fear*, p. 13. Andren opens his discussion of national security in much the same fashion. Andren, "In Search of Security," p. 217.

21. Ullman, "Redefining Security," pp. 130–31.

21. Peter Katzenstein, *West Germany's Internal Security Policy: State and Violence in the 1970s and 1980s* (Ithaca, N.Y.: Cornell Western Societies Papers, 1990).

23. John Herz, *Political Realism and Political Idealism* (Chicago: University of Chicago Press, 1951); Buzan, *People, States, and Fear*, 2d ed., ch. 8.

24. Report of the Independent Commission on Disarmament and Security Issues, *Common Security* (London: Pan, 1982).

25. Haftendorn, drawing inspiration from Hedley Bull, suggests that three schools of thought are identifiable, each associated with a political philosophical "founding father": national security (Hobbes), international security (Grotius), and global security (Kant) Haftendorn, "The Security Puzzle," pp. 4–13.

26. In his *Neutral Democracies and the New Cold War*, Sundelius presents an innovative menu of strategies potentially available to small states (and large ones) in dealing with their security problems See also P. Bröms, *Environmental Security Regimes: A Critical Approach* (Stockholm: Swedish Institute for International Affairs, 1995).

27. Cited in Brown, "Climate, Ecology, and International Security," p. 521.

28. Cited in Brown, "Climate, Ecology, and International Security," p. 522.

29. Report of the Commission on Global Governance, *Our Global Neighborhood* (Oxford: Oxford University Press, 1995).

30. For a collection of statements linking environment and security by US. officials, see Environmental Change and Security Project Report, "Environment and Security Debates: An Introduction," pp. 47–58; and Al Gore, *Earth in the Balance: Ecology and the Human Spirit* (New York: Houghton-Mifflin, 1992).

31. President William Clinton, *National Security Strategy of the United States: 1994–1995* (London: Brassey, 1993/1995).

32. Ole Waever, "Securitization and Desecuritization," in Ronnie Lipschutz, ed, *On Security* (New York: Columbia University Press, 1995), pp. 46–86.

33. For example, Goldmann describes national security in the following terms: "National security policy will, one can say, be geared towards minimizing the price which must be paid for autonomy in the form of warfare and war risks" [author's translation]. K. Goldmann, *Det Internationella Systemet* (Stockholm: Aldus, 1978), p. 54.

34. Stephen Walt, "The Renaissance of Security Studies," *International Studies Quarterly* vol. 35, no. 2 (1991), p. 212.

35. Walt credits Joseph Nye and Sean Lynn-Jones with this, but they are actually more receptive than Walt to other issues as security threats: "The central questions are concerned with international violence, but there are also other threats to the security of states" Joseph Nye and Sean Lynn-Jones, "International Security Studies," p. 6.

36. Walt, "The Renaissance of Security Studies," p. 213.

37. Walt, "The Renaissance of Security Studies," p. 227.

38. Crawford places study of the "relationship between environment and security" at the very heart of her proposed research agenda for security studies Crawford, "Once and Future Security Studies," p. 308.

39. For example, see M D'Antonio, *Atomic Harvest: Hanford and the Lethal Toll of America's Nuclear Arsenal* (New York: Crown Publishers, 1993); and Murray Feshbach, *Ecological Disaster: Cleaning up the Legacy of the Soviet Regime* (New York: Twentieth Century Press, 1995).

40. For a rigorous conceptualization of influence attempts, see David Baldwin, *Economic Statecraft* (Princeton, N.J.: Princeton University Press, 1985).

41. For documentation of the manipulation of food supplies for political purposes in Africa, see J Macrae and A. Zwi, "Food as an Instrument of War in Contemporary African Famines: A Review of the Evidence," *Disasters*, vol. 16, no. 4 (1992), pp. 299–321.

42. T. M. Hawley, *Against the Fires of Hell: The Environmental Disaster of the Gulf War* (New York: Harcourt Brace Javonovich, 1992), p. 8.

43. Walt, "The Renaissance of Security Studies," p. 227.

44. For an analysis of international laws related to environmental destruction as a method of warfare, see J. Nordenfelt, "Environmental Destruction as Method of Warfare" in Hjort af Ornäs and Lodgaard, eds., *The Environment and International Security* (Uppsala: PRIO/Uppsala University Program on Environment and International Security, 1992), pp. 23–34.

45. For example, the Roman decision to sow the Carthaginian agricultural land with salt after the Punic Wars may be interpreted in these terms.

46. The proliferation of high-risk technologies such as nuclear power and the chemical industry creates new kinds of problems For example, the 1981 Israeli bombing of the Iraqi reactor at Osiraq highlights the vulnerability of such facilities to military attack. For discussion, see: Bennett Ramberg, "Attacks on Nuclear Reactors: The Implications of Israel's Strike on Osiraq," *Political Science Quarterly* vol. 97, no. 4 (1982–83), pp. 653–62.

47. For extended analysis of the environmental impacts of warfare, see Arthur Westing, *Explosive Remnants of War: Mitigating the Environmental*

Effects (London: Taylor and Francis, 1985) An estimated ten million landmines were left in Afghanistan after the Soviet intervention. Senator Patrick Leahy, "Landmine Moratorium: A Strategy for Stronger International Limits," *Arms Control Today* (January–February 1993), pp. 11–14.

48. For example, see Brown, "Climate, Ecology, and International Security."

49. Thomas Homer-Dixon, "On the Threshold: Environmental Changes as Causes of Acute Conflict," *International Security*, vol. 16, no. 2 (1991), pp. 106–14.

50. S. Lodgaard, "Environment, Confidence Building, and Security," in Hjort af Ornäs and Lodgaard, eds., *The Environment and International Security* (Uppsala: PRIO/Uppsala University Program on Environment and International Security, 1992), pp. 11–22.

51. Walt, "The Renaissance of Security Studies," p. 212.

52. Buzan, *People, States and Fear*, 2d ed.

53. For further discussion, see the chapters by Butts and Deibert in this volume.

54. Deudney, "The Case Against Linking Environmental Degradation and National Security," *Millennium*; Deudney's argument is extended in chapter 8 of this volume.

55. Matthias Finger, "The Military, the Nation-State and the Environment," *The Ecologist* vol. 21, no. 5 (September/October 1991), pp. 220–25; and Ken Conca, "In the Name of Sustainability: Peace Studies and Environmental Discourse," *Peace and Change* vol. 19, no. 2 (1994).

56. Thomas Homer-Dixon, et al., "Environmental Change and Violent Conflict," *Scientific American* (February 1993), pp. 38–45; and Thomas Homer-Dixon, "Environmental Scarcities and Violent Conflict: Evidence from Cases," *International Security* vol. 19 (Summer 1994), pp. 5–40.

57. Buzan, *People, States and Fear*, p. 245. The second edition of Buzan's book reveals an increased emphasis on the centrality of the state in the posited network of security relationships across levels. See, for instance, *Buzan, People, States and Fear*, 2d ed., p. 328.

58. Ken Booth, "Security in Anarchy: Utopian Realism in Theory and Practice," *International Affairs* vol. 67, no. 3 (1991), pp. 527–45.

59. Deudney, "The Case Against Linking Environmental Degradation and National Security," p. 469.

60. Levy, "Is the Environment a National Security Issue?"

61. Levy defines the term *environment* as referring to "issues involving biological or physical systems characterized by significant ecological feedbacks or by their importance to the sustenance of human life," p. 39. For alternative definitions of environment, see P. Bröms, *Environmental Security Regimes: A Critical Approach* (Stockholm: Swedish Institute for International Affairs, 1995); and M. Tennberg, "Risky Business: Defining the Concept of Environmental Security," *Cooperation and Conflict* vol. 30, no. 3 (1995), pp. 239–58. I specify the types of environmental values most likely to be perceived as security threats in the following, admittedly anthropocentric, terms: "threats to human health (such as highly contagious bacterial or viral diseases), threats to essential resources which support life in human ecosystems (e.g., air, water supply, and food production), and threats to the integrity of valued non-human eco-systems and local/global biological diversity." Eric Stern, "High Politics and Crisis: Military and Environmental Dimensions of Security," in Hjort af Ornäs and Lodgaard, eds., *The Environment and International Security* (Uppsala: PRIO/Uppsala University Program on Environment and International Security, 1992), p. 90, fn.13.

62. Levy, "Is the Environment a National Security Issue," p. 60.

63. Levy, "Is the Environment a National Security Issue," p. 61.

64. Levy, "Is the Environment a National Security Issue," p. 50.

65. Levy, "Is the Environment a National Security Issue," p. 54.

66. Levy, "Is the Environment a National Security Issue," p. 36.

67. For example, see Mische, "Ecological Security and the Need to Reconceptualize Sovereignty."

68. Alexander L George, *Bridging the Gap: Theory and Practice in Foreign Policy* (Washington, D.C.: U.S. Institute of Peace Press, 1993), p. 27.

69. John Steinbruner, *The Cybernetic Theory of Decision* (Princeton, N.J.: Princeton University Press, 1974); and J. March and J. Olsen, *Rediscovering Institutions: The Organizational Basis of Politics* (New York: Free Press, 1989).

70. For example, see Robert Jervis, *Perception and Misperception in International Politics* (Princeton, N.J.: Princeton University Press, 1976), pp. 128–41.

71. Irving Janis and L. Mann, *Decisionmaking: A Psychological Analysis of Conflict, Choice, and Commitment*, (New York: Free Press, 1977); and George, *Presidential Decisionmaking in Foreign Policy*.

72. Y. Vertzberger, *The World in Their Minds* (Palo Alto, Calif.: Stanford University Press, 1991), pp. 138–39.

73. R. Kenney, *Value-Focused Thinking: A Path to Creative Decisionmaking* (Cambridge, Mass.: Harvard University Press, 1992).

74. Kenney, *Value-Focused Thinking: A Path to Creative Decisionmaking*,

p. ix.

75. See: Dalby, "Security, Modernity, Ecology," pp. 113–14. A good example of a phenomenon that could develop into an environmental security dilemma is the export of toxic waste from developed to underdeveloped countries. The attempt on the part of the developed country to protect its own population creates a risk to population of the underdeveloped country.

76. The concept of "sustainable development" is another good example of a concept that facilitates constructive engagement of serious (and previously largely overlooked) value conflict That concept addresses the potential conflict between values associated with conservation/environmental protection on the one hand, and rapid economic growth/development on the other. For a conceptual and empirical analysis of sustainable development, see B. Fritsch, "On the Way to Ecologically Sustainable Economic Growth," *International Political Science Review* vol. 16, no. 4 (1995), pp. 361–74.

77. John Dewey, *Experience and Nature* (Lassalle, Ill.: Open Court, 1929/1989), p. xiii.

7

Threats from the South?

Geopolitics, Equity, and Environmental Security

Simon Dalby

"Environmental Security"

Since the end of the cold war there has been much policy and academic debate about how to reformulate concepts of national and international security now that the superpower rivalry no longer dominates discussions of global politics. Some key policy discussions have linked national and international security to questions about the potential future dangers of social and political disruptions resulting from environmental degradation. The key themes were popularized in 1989 by Norman Myers, Jessica Tuchman Mathews,

155

and Michael Renner.[1] They all include elements of security rethought in much more comprehensive fashion to incorporate the environment as a framework for human security in general. Many policymakers and academics now discuss these matters in terms of "environmental security."[2]

The argument that global environmental degradation is a common security threat to all humanity has very considerable rhetorical force. It is a term that both emphasizes the need to take environmental issues seriously as a matter of both national and international politics and simultaneously suggests an urgency in dealing with global problems portrayed as threatening to specific societies. As Myers, Mathews, and Renner argued, it clearly points to an urgent need to consider the dangers of global environmental degradation, potential resource constraints, and dramatic demographic changes. It raises concern about the immediate practical consequences and the political perils that are likely to occur as a result of increasing renewable resource shortages and environmental difficulties resulting from deforestation, desertification, soil erosion, industrial pollution, and the potential future disruptions caused by climate change. The use of the term *security* suggests the need to counter practical threats that endanger all people as well as states. In the context of "ozone holes" and global climate change, "environment" is often understood to apply to worldwide phenomena. The term *global* has also often been applied to the related impending difficulties that apparently require widespread international cooperation to effectively tackle.

All these arguments rely on a comprehensive notion of security, one applied to long-term and large-scale thinking that transcends the narrow focus of Cold War "geopolitical" national security thinking with its focus only on political and military threats to states and their dominant economic interests. The term *environmental security* has become part of a framework for advocating new conceptions of how the international political order should be understood and what normative aspirations are appropriate in the post–Cold War period. The 1994 United Nations Development Program *Human Development Report* listed environmental security as one of the seven components of its formulation of global human security.[3] In 1995 the Commission on Global Governance followed this line of argument to add planetary security to its reformulation of the post–Cold War security agenda.[4] In 1997 the Clinton administration adopted environmental themes as a focus for its foreign policy, a NATO research project was underway, and the Woodrow Wilson center had undertaken a major project to examine the environmental dimensions of foreign policy.[5]

Environmental Security: An Ambiguous Concept

But like many of the terms used in contemporary politics, on closer inspection environmental security or "ecological security" as it is sometimes rendered appears to be an ambiguous phrase. The terms of the discussion are rarely precisely defined, and consequently the links between environment and specific entities that may be rendered insecure are sometimes less than clear. Discussions of human security or global environmental security are often very diffuse as a consequence. Of course, this is often what makes such language so politically useful. Academic critics of the policy discussion argue that the debate is so vague as to make it impossible to render the matters in the discussion intelligible with enough precision to allow for a formal academic research agenda to confirm or refute many of its contentions.[6] Thomas Homer-Dixon, one of the most widely cited authorities on environmental degradation and violent conflict, usually simply avoids discussing these matters specifically in terms of environmental security.[7]

But there is much more than an academic research agenda involved in these discussions. The debate about environmental security is about how politics will be rethought and policy reoriented after the Cold War. Conflating this and the academic agenda often simply causes confusion.[8] The use of the term by the U.N. Development Program and the Commission on Global Governance suggests clearly a political exercise about whose issues are part of the international policy agenda. It is also to be expected that policymakers and institutions with specific political interests will attempt to co-opt advocates of positions and arguments that they find useful. The military can sometimes be "green" when it suits its institutional purpose, and intelligence agencies may also seek roles in monitoring environmental trends.[9]

In this process it is not surprising that broad generalizations proliferate along with assumptions of common global interests among all peoples. But global or universal political claims often have a nasty habit of turning out to be parochial concerns dressed up in universalist garb to justify much narrower political interests. This chapter argues that much of the policy literature linking environmental issues and security (broadly defined) is in danger of overlooking important political issues unless analysts are alert to the persistent dangers of the traditional ethnocentric and geopolitical assumptions in Anglo-American security thinking.[10] Security thinking is only partly an academic discourse, it is, as recent analysts have made clear, much more importantly part of the process of international politics and the formation of American foreign policy in particular.[11] This suggests that if old ideas of security are added to new concerns about environment the policy results may not be anything like what the original advocates of environmental security had in mind.

There are a number of very compelling arguments already in print that suggest some considerable difficulties with the positing of environmental security as a "progressive" political discourse.[12] While the argument in this chapter acknowledges the efficacy of the case against environmental security as a policy focus, the point of departure takes seriously the political desire to fundamentally rethink the whole concept of security as a strategy to reorient political thinking and to extend definitions of security, of who and what should be rendered secure, and also who should be the political agents providing these new forms of security. While these "progressive" ideas may be a minority concern on the political landscape, they are interesting both because they shed light on conventional thinking and because they suggest possibilities for rethinking conventional state-dominated political concepts and practices.

In particular the assumptions that states really do operate in the interests of their national population needs to be reexamined. Military organizations are not necessarily in the business solely of protecting domestic populations from external threats. As the persistence of at least some military dictatorships, and the numerous intrastate conflicts of the 1990s indicate, they often endanger "domestic" populations more than they protect against external intrusions. In addition, the common assumptions that economic development as conventionally practiced is necessarily going to provide either directly, or indirectly though state agencies, security for populations in underdeveloped parts of the world is also dubious. It is important to remember that the premise of the term *sustainable development* is that conventional development is not environmentally sustainable. Finally, in considering the questions of environmental security at the large scale it is also important to keep in mind the international flows of resources and wealth in the global economy, matters that conventional international relations thinking often obscures by its focus solely on states as political actors.[13]

Examining the environmental security literature in the light of these arguments suggests some fundamental critiques of the global order that has been in part "secured" by a geopolitical understanding of American national security through the Cold War period. Further, and of primary concern to this chapter, this is especially clear when the environmental security discourse is confronted by the issues of global justice in terms of a "Southern critique" of the double standards and inequities in some high profile Northern environmentalist policies. These lines of argument suggest that unless the term *security* is understood in substantially different ways in the new circumstances of the 1990s, it is quite possible that "environmental security" may turn out to be just another addition to the traditional Cold War American policy concerns with national and international security that usually operated to maintain the global political and economic status quo.[14]

To many of its Southern critics, this world order is precisely what has rendered so many people insecure in so many ways. According to the 1994 *World Development Report*, during the period from 1960 to 1991 the ratio of the percentage of the world's wealth held by the richest 20 percent of the world's population to that by the poorest 20 percent went from 30:1 to 61:1.[15] Most of the wars for the last half century have occurred on the territories of the underdeveloped states, and continue to be fought with weapons supplied from the arms industries of Northern states.[16] And despite numerous claims that states are formally equal, contemporary processes of globalization work in many ways to the disadvantage of weaker and poorer underdeveloped states.[17]

Global Security and Geopolitics

In the North Atlantic Treaty Organization (NATO) countries, Cold War national security policy emphasized the geopolitical containment of the Soviet Union, principally through the strategy of military deterrence to "contain" the geographical expansion of Soviet influence. But these geopolitical strategies of dividing up and controlling the world also deterred other challenges to the American-supported political and economic order around the globe.[18] American development aid, security assistance, and military alliances involved both active foreign interventions and some degree of control of the global political economy to ensure the flow of key resources to the United States, Europe, and Japan.[19] Much of the world's wealth and technology is now either in the North, or controlled by Northern-based global corporations.[20] When its focus is on resource supply sources in the South, the discourse of environmental security can easily extend the explicitly geopolitical understandings of security that were widespread during the Cold War to maintain the current global division of wealth.

Concern about environmental security, particularly when it entails policies of limiting the use of resources, slowing population growth, and curtailing specific economic activities in the South, can thus easily be criticized as just one more political tactic on the part of those in the North who wish to maintain their control over global politics and resource flows. This "Southern critique," in numerous statements by Southern delegates at the 1992 United Nations Conference on Environment and Development (UNCED) meetings in Rio de Janeiro, asserts that the North is using fears of environmental disaster as a weapon to limit Southern development, or to control its political arrangements to ensure that Northern economic interests continue to have access to the traditional cheap Southern labor and resources.[21]

Thus, as is the case with questions of security in general, the most interesting questions confronting any discussion of the themes of environmental security concern what is being rendered secure, by what means and for whom.[22] The converse is to inquire closely as to what the "threat" is to this "security." Both the Southern critique of global inequities and the geopolitical presuppositions in contemporary security thinking in many Northern states raise these questions. In doing so they focus on the necessity to drastically rethink the term *security* if it is to be attached to the term *environment* in future debates about global politics and the possibilities of sustainable livelihoods.

Security has long been understood as the protection of national sovereignty and the political integrity of the state. In general it is understood in the geopolitical terms of insides and outsides, protecting the internal "us" from the threatening external "other."[23] The advantage of this formulation is the ease with which domestic political resources can be mobilized to support state efforts to "protect" the internal society from external threat. Simplistic ethnocentric images can be marshaled to demonize external enemies and prepare to combat external dangers. But the problem with these geopolitical specifications of security is that the resulting policies are nearly always understood as reactive in the sense of policymaking responding to supposedly completely external independent contingencies.[24] Security in the Cold War was understood in terms of external threats coming from some place beyond the sphere of domestic political action and control. The response to these external threats was military force applied to counter the threat. As formulated most clearly in the doctrine of deterrence they tended to be negative and violent attempts to assert control over the "external" global polity.

The focus on the natural environment in environmental security thinking often meshes with more traditional understandings of the security environment as the constellation of external political and military forces facing a particular state. This understanding implies that traditional security policies are appropriate in the new circumstances of global environmental "threats." This approach also implies that military solutions are available to resolve at least some of these difficulties, an assumption that is often very wide of the mark if detailed examinations of the root causes of many environmental difficulties are undertaken. The danger here is that environmental matters will be militarized rather than dealt with by more appropriate policies that are sensitive to economic factors and the local context.[25]

While the politics of deterrence remained the paramount concern in American security planning during the Cold War, other matters such as threats to energy supplies after the OPEC experience of the 1970s, threats to domestic economic performance partly due to growing imports and as a response to growing trade deficits with Japan in the 1980s, and the threat of imported illegal drugs had been added into discussions of U.S. national security.[26] The geopolitical assumptions in security thinking reduce any sense of historic responsibility or prior involvement with the foreign places that produce threats.[27]

By focusing on environmental matters in terms of security, attention is also diverted away from "internal" matters of consumption and resource usage in the developed states where a small minority of the world's population consumes a disproportionate amount of world resources, and produces huge quantities of pollution and greenhouse gases. The "external" threats to this consumption are defined as the problem and the focus is shifted to possible political disruptions from environmental refugees and migrations caused by external environmental disruptions and political turmoil. Apocalyptic visions of political disintegration in Africa and elsewhere are often portrayed as events happening in areas remote from Northern states, and as being unrelated to global patterns of poverty and underdevelopment.[28]

Environmental security understood in these traditional terms could thus be used to support the "global managerialist" ambitions of some Northern planners, justified in part by the apparent need to control the international environment in the unpredictable period after the collapse of the Cold War geopolitical order.[29] The logic of such thinking is particularly clear in a speech given by Manfred Worner, then NATO's secretary general, shortly after the demise of the

Cold War. In summarizing the role of NATO he declared that Western states are increasingly concerned with stability at a global scale.

> In the first place, the Allies recognize that, as the direct military threat recedes, security no longer needs to involve a staving off of immediate military peril; but that the underlying security task is the long-term provision against possible security risks— instability, uncertainties, the possibilities of a new aggressiveness and future force generation from those who now appear less bellicose.[30]

But beyond these generalities and general fears of the unknown in the future, the geopolitical division between "the West and the rest" becomes much clearer further on in the speech:

> As the declaration at the Summit meeting in 1989 has already foreshadowed, we are increasingly broadening our security notion to include both new military threats and the non-military challenges that come to us from the Third World. The immense conflict potential that is building up in Third World countries, characterized by growing wealth differentials, an exploding demography, climate shifts and the prospect of environmental disaster, combined with the resource conflicts of the future, cannot be left out of our security calculations, no matter how we translate our broader analysis in operational aspects in the longer term.[31]

But Worner is clear that the management of the threats to global security will clearly come from the West and Japan. They will provide the wealth and supervise the management of problems in the twenty-first century, hopefully in increasing concert with the powers of Eastern Europe. Because the NATO states and Japan between them generate more than half the world's wealth:

> That places a tremendous responsibility upon them for managing their own mutual relationships harmoniously as a precondition for managing the larger affairs of our world. The secret of security in the future is that it is increasingly indivisible and that we all share the responsibilities for its management.[32]

This NATO understanding of the post–Cold War world is clearly one of the persistence of Northern institutions as the core political

arrangements from which the rest of the world can be "managed."
The theme of a select few managing the world's affairs is clear.
There is no place in these arguments for a powerful United Nations or
an international dialogue considering the interests of the less powerful members of the world community. Environmental crises are seen
as emanating from the Third World rather than as the result of industrial states' economic activities with their related pollution and ecological disruptions. Questions of global justice are nowhere to be found in
these scenarios. Maintaining the status quo order is much more important. But there is no room here for nonindustrialized states to participate in shaping that global political order. Neither are there possibilities for new economic ventures designed to rectify the growing
gap in wealth between the North and the South. The crisis must be
managed, and the redistribution of wealth, power, and resources are
simply not considered. Once again the world is "orientalized"—divided
in this case into the North and the South, with the North making the
decisions and the rest of the world being managed, with or without
their consent. All this is done in the name of maintaining security.

Geopolitics, Equity and Security

The point here is not that the North has operated in this fashion in
the last few years. Internal rivalries, economic problems, and preoc-
cupations with Bosnia and the Middle East have in any case focused
attention elsewhere. Rather the point is that the addition of environment as a theme to the security agenda of NATO, without a drastic
rethinking of the term *security* and who and what is to be secured,
leads easily to the traditional geopolitical formulations of policy in
terms of external threats requiring political management, or in the
last resort, military intervention to maintain a political order that is
dominated by the rich and powerful Northern states.

As the following sections make clear, similar possibilities run
through the formulation of many other aspects of the themes that
come under the rubric of environmental security. The key point is
that "environment" cannot be simply grafted onto "security" without
the politics of how this is done being very carefully examined.

Global Resource Use and the Object of Security

Through the Cold War period repeated concerns were raised by security analysts in the United States about the vulnerability of the

American economy, particularly its military industrial production base, to external disruptions of supplies of raw materials.[33] In the 1970s resource access arguments were made in terms of the growth of Soviet geopolitical threats to Western supply "lifelines."[34] This in part triggered American intervention in various places to secure trade routes and establish bases to defend shipping routes. The most obvious recent case is the role of the United States in the Gulf during the Iran-Iraq war in the 1980s in attempting to secure access to oil supplies against Iranian or Iraqi threats to disrupt either production or tanker traffic. The industrial states are heavily dependent on external supplies of resources, particularly oil. Western geopolitical strategies for resource access take for granted that industrial states have a right to these supplies whatever claims to sovereignty governments may try to make, and that military interventions, to ensure that supplies flow, are justified.

The persistence of the focus on these matters in terms of geopolitics concentrates attention on external matters rather than internal changes as the response to changing international circumstances. Constructing the Gulf solely in terms of a geopolitical arena of importance because of its supplies of petroleum ignores the demand side of the energy situation. As Amory and Hunter Lovins have pointed out repeatedly, the easiest option for U.S. foreign policy in the Gulf is energy-efficient automobiles and roof insulation in North American buildings.[35] This is one of the clearest lessons of the 1991 Gulf War. The United States, as well as Japan and Europe, will remain vulnerable to political developments in the Gulf if they do not undertake serious measures to reduce oil consumption. "If the United States wants to be serious about reducing the degree of its political vulnerability to domestic and international upheaval in the Middle East, then it must raise taxes on gasoline, motor fuels, petroleum products or carbon emissions."[36] This point makes the political dilemmas of environmental security thinking very clear.

In its Cold War manifestations, security referred to the maintenance of the political order of the Western world, or more broadly, modernity. The particular societal organization of what is now probably better termed "the North," with its (post) industrial economy, is based on cheap fossil fuels. The automobile age is the fossil fuel age. While the automobile may no longer be its leading innovative technology, the economic system spawned on the basis of suburbanization and rapid transportation has spread to many parts of the globe. The mass consumption of resources, and in particular fossil fuels, is the sine qua non of this economic system.

But it is precisely this use of fossil fuels that is changing the planet's atmosphere, with as yet uncertain consequences on the global climate. Insofar as maintaining the integrity of the ecosphere is considered a part of the agenda for a policy on environmental security then it is precisely the social system that "national security" has been supposedly protecting that is the source of the problem. This question of security understood in resource and economic terms illustrates the importance of understanding security in relation to specific political identities. Insofar as what is being rendered secure is a version of modernity dependent on the massive use of fossil fuels, the environment must necessarily be rendered insecure. If the environment is to be "secured," as in preserved and protected, then Northern consumption patterns will have to be modified. But it has long been the "right" to maintain Northern lifestyles and the power structure that underlies them that has been the object of the geopolitically formulated national security policies of Northern states.

"Global" Issues and Specific Responsibilities

The questions of global equity and issues of intergenerational responsibilities are also connected with environmental security and the crucial question of who and exactly what are to be secured. It is often claimed that climate change is a global problem needing global solutions. This was the rationale for the climate convention negotiations that are part of the 1992 UNCED agreements.[37] But the Southern critique points out that the industrialized states and their large use of fossil fuels in particular are responsible for a large proportion of the total greenhouse gas emissions. These states also have the wealth to tackle the problems. Of course many underdeveloped states produce greenhouse gases, but their total output is far lower on a per-capita basis. Furthermore, as Anil Agarwal and Sunita Narain point out, Southern emissions, and the contribution of tropical deforestation to carbon dioxide and methane production, are often a result of "survival" activities, rather than "luxury" emissions that come from Northern consumer lifestyles.[38] Methane produced from a peasant plot in an agricultural society is much more immediately necessary to human welfare than an equivalent discharge from an industrial concern in a Northern state producing some luxury consumer item. Likewise they argue that it is inappropriate to blame Algeria for having large methane emissions from its natural gas facilities, because the gas is produced for export and used by Europeans, not by

Algerians. A similar argument could be made about tropical deforestation to serve Northern timber demands. Because of these equity concerns it is not surprising that many in the South object to being asked to pay the price for "fixing" the problems that analysts in the North define as "global."

Particularly important here are the common geographical assumptions that global environmental degradation is a problem that is widely recognized as being a matter of common interest, caused in similar ways in very different places and contexts, and hence requiring global solutions.[39] Such assumptions often obscure the specific environmental insecurities that people experience in particular places as a result of global political and economic forces. This is perhaps clearest in the case of the other highly prominent global atmospheric problem, stratospheric ozone depletion, which largely results from Northern industrial activities, particularly the production of chlorofluorocarbons (CFCs). But as Vandana Shiva very bluntly puts it:

> That such substances as CFCs are produced by particular companies in particular plants is totally ignored when ozone depletion becomes transformed into a "global" environmental problem. The producers of CFCs are apparently blameless and the blame laid instead on the potential use of refrigerators and air-conditioners by millions of people in India and China. Through a shift from present to future, the North gains a new political space in which to control the South. "Global" concerns thus create the moral base for green imperialism.[40]

The logic of this argument suggests that blaming Southern states for their reluctance to agree to international conventions in the absence of some financial compensation for foregone development opportunities, replays the geopolitical game of assigning blame and responsibility to external others in ways that obscures "domestic" responsibilities.

These equity concerns are central to the negotiation of global environmental agreements, and ignoring them will only delay attempts to solve environmental problems. Because the Northern countries consume disproportionately large amounts of the globe's resources, produce most of the CFCs, and most greenhouse gases, then clearly, so the Southern critique suggests, the bulk of initiatives will need to be taken by the North. This is clearly the lesson to be learned from the international agreements that have been worked out to deal with the stratospheric ozone-depletion problem. After contentious negotiations, the international protocols limiting CFC production did eventually incorporate financial and technological transfers to poorer states to mitigate their anticipated economic losses as a result of the need to abandon CFC production and substitute other processes.[41] If thinking about environmental matters in the misleading terms of "global security" obscures these equity questions then initiating change in individual countries and fashioning global agreements will be much more difficult.

Global Commons or National Sovereignty?

Whatever their source, ecological problems like global atmospheric change and ozone depletion are clearly global in their consequences, because they involve physical changes to the ecosphere in its totality. The atmosphere, like the oceans or Antarctica, is a global common property resource which no single state owns or has jurisdiction over, but from which all gain some benefit. In the case of a resource in limited supply, a clearly understood and administered agreement on the allocation of the resource can ensure that benefits continue to be available for all. The problem with common property renewable resources is that in the absence of such an agreement, each user is likely to use as much as they individually want, which may be more, collectively, than the resource can supply. Overfishing on the high seas, which has caused fish populations to sharply decrease in many places, is a classic example of this process.[42]

The atmosphere is a commons, but are the remaining forests and their intrinsic biodiversity also a commons? In the biodiversity negotiations leading up to the Earth Summit in 1992, and in discussions about other forest agreements and conservation plans, many Northern environmentalists have made suggestions that the remaining intact forests on Earth, mainly in tropical and underdeveloped areas, should be treated as global commons to ensure the survival of global biodiversity.[43] The richest concentrations of the immense complexity of terrestrial species on the planet are contained in the old tropical forests. Thus the argument suggests that these are unique resources for the whole planet and it is necessary to have them administered in a way that ensures all future generations may benefit from any medical or commercial products that they are found to contain. In addition the argument is often made that these forests are crucial carbon sinks for the excess carbon dioxide in the atmosphere. Their preservation is thus deemed essential to stabilize climate and efforts to commonize the forests have been employed to justify intrusive and extensive international regulation of tropical forests.

But any attempt to make these forests a global commons, administered by international treaty, runs directly into the question of the political remit of sovereignty.[44] Unlike the atmosphere and oceans, states have sovereign claims to the land on which the forests grow. Although current biodiversity losses are more concentrated in the tropical rain forests, many governments of these states argue that these forests and their biological fecundity are the sole possession of their peoples, rather than the common heritage of humanity. The Brazilian objections to foreign interference in the development of the Amazon is a clear example of states clinging to their sovereign rights to make decisions about resource and environment issues.[45]

The reasons for this assertion of sovereignty are obvious. "Development" in the Brazilian Amazon is understood by many there as essential to the future security of the state, both because it facilitates resource use and allows easy military access to remote regions. Although conventional development strategies have not alleviated the lot of the poor or protected indigenous lands, and have indeed accelerated their migration into the vulnerable tropical forests, this kind of geopolitical thinking has greatly influenced the evolution of Amazonian policy.[46] Development understood in these terms leads Southern countries to see environmental forest protection as threats to the development that would enrich the state and provide jobs for the population. International environmental agreements that constrain development are then seen as a threat to the national security of the state.[47]

But Northern demands for common administration of the remaining forests appear unjust when it is remembered that the Northern states became wealthy through a process of nearly completely deforesting much of their territory before concerns about the global impacts of such devastation were on the political agenda. In addition Northern pharmaceutical, seed, and agricultural companies often patent chemical products derived from species found in the tropical forests, usually without any royalties being paid to the state of origin of the species involved, thus depriving underdeveloped states of the value of their resources.[48] The bulk of greenhouse gases come from Northern consumption of fossil fuels and not from Southern deforestation, so it is obviously unfair, the argument goes, to ask the poor states in the South not to cut their forests so that Northern carbon dioxide emissions will not cause climate change. Combined, these environmental agendas can easily be seen through Southern eyes as a plan to maintain the South in poverty by denying it the right to follow the Northern route to development and prosperity. Or, in more sophisticated accounts, the global disparities are seen as the primary obstacle to building sustainable modes of living in both North and South.[49] Clearly, the structural inequities in global economic and political power also put many Southern states at a disadvantage in negotiating international environmental agreements.[50]

A simple illustration reversing the geopolitical situation might help clarify this reasoning. A Southern response using the same logic of claiming a global commons for essential resources might suggest that the American grain belt be made a global common resource.[51] Given its productivity, and the clear need in many parts of the world for much improved nutrition, the "commons" reasoning can lead to the conclusion that American claims to national sovereignty should be overridden by the needs of the world's population for food. Thus the logic might suggest that an international regime should be established to ensure that American grain is made available wherever it is needed around the planet regardless of what the American government may think of this arrangement. The objections to such a suggestion are likely to be considerable in Washington. While the analogy is imperfect, the heuristic exercise is useful if it clarifies the ethnocentric limitations in some of the conventional Northern environmental thinking.

Coerced Conservation

Advocates of environmental security as a policy framework have sometimes suggested a new role for the military in the preservation of environmental systems.[52] If military establishments are seen as upholding their respective national interests, and if the national interests of many states now include resource conservation and environmental protection, then it supposedly follows that military institutions should be used to protect environments and facilitate resource development. More specifically the military could act as "super park wardens," enforcing bans on hunting and using sophisticated intelligence-gathering technology to police ecological reserves. These proposals parallel others to "convert" the post–Cold War military of many states into more contemporarily useful institutions, and simultaneously the apparent need under the logic of environmental security to use drastic measures to protect the global environment.

But in many parts of the world the military plays a role quite different from that in the Northern democracies. In many Southern states the military is a disproportionately powerful social institution with a direct stake in the economy. Many Southern states also have ethnic and elite competitions for power which involve the military in violent internal conflicts. Far from being an agency outside formal politics, internal political struggles for power often involve the military in political repression and violence, making their populations vulnerable to, rather than protecting them from, violence.[53] In these circumstances using the military is rarely adequate to either protect "nature" or deal with the social problems that endanger the ecosystems that are supposedly being protected. Use of the military to "coerce conservation" fails to deal with the poverty and lack of resources that many in the South face.[54] Imposing conservation efforts without local consent and support is unlikely to be successful in the long run because it fails to alleviate the poverty that often leads to environmental degradation. Poachers and illegal logging activities are likely to continue so long as legal economic alternatives are not available. But with the support of local communities, who have a clear economic interest in maintaining specific ecosystems, the need for a military type approach or "imposed conservation" is often eliminated.[55]

The dangers of coercive conservation are clear when the question of what is being conserved (secured?) where and for whom is asked. Parks are the administrative model used by most Northern governments to protect isolated environments that are deemed either scenic or ecologically valuable. But transferring this model to Southern states lacking competent administration and effective democratic accountability, in situations where title to land is often disputed by indigenous or nomadic people, may be more of a license to coerce than to conserve. Nature reserves are often a legacy of colonial administrations that set them aside to conserve game animals so that they could be hunted by colonial settlers or foreign tourists rather than used for the benefit of indigenous inhabitants who were relegated to the outlaw category of poachers.[56] "Ecotourism" allows for the protection of many large animals but it may still do so at the cost of removing local people from traditional grazing lands and producing an artificial environment in which animals can be viewed.

When military organizations play important roles in resource management, adding the enhanced legitimacy of global environmental protection, and military resources to conservation agencies to support their policies, may only worsen violence where central government policy collides with local customary practice. The specific roles of military governments in controlling resource developments in their own interests often directly conflict with the interests of the local population. Overlooking this ethnocentric assumption about the social role of the military is yet another failure to think locally while also trying to think globally. It also suggests that many environmental issues might be dealt with better by following a policy of "desecuritization" where matters are removed from the jurisdiction of the military and dealt with by other social agencies and actors.[57]

Global Population

Another arena where environmental security thinking relates to contemporary political matters is in the discussion of global population.[58] Unlike most industrial countries the populations of many Southern countries have not completed the process of the "demographic transition" in which death rates first fall rapidly as a result of improved nutrition, sanitation, and disease control, and then birth rates fall as families adapt to new economic and cultural situations.[59] Thus the threat to security can be portrayed as the fecundity of the poor, a fecundity that it can then be argued obviously needs to be tackled by state interventions and government controls.

Unfortunately the record of imposed population policies in the last few decades in various parts of the world, particularly China's one-child family program, and India's experiments with compulsory sterilization, indicate the coercive implications of demographic concerns understood in terms of national security. Many argue that the alternative to government-coerced fertility reductions is women's empowerment linked to basic health care and contraceptive availability. But this challenges the male-dominated status quo in many places, and the "security" of those who benefit from these social arrangements. Nonetheless the results of the 1994 United Nations population conference held in Cairo, in which the final program of action stressed the importance of female empowerment, health care and contraceptive availability, as well as the right of all people to make their own decisions about reproduction, indicate that women's organizations are finally getting these themes firmly on the agenda of international political discussion.

Some analysts forcefully argue that the increasing numbers of refugees and economic migrants may be a potential threat to political order and stability in the West.[60] If the planet cannot support its growing population with Northern middle-class lifestyles to which all aspire, then a crisis is likely to emerge as growing populations of the Southern states clamor for the wealth controlled by a shrinking percentage of the world's population in the Northern states. Migration may upset the political organization in Northern states causing new forms of insecurity among the aging populations of postdemographic transition societies.[61] Some in effect make a case for dividing the world on the basis of rich and poor, if not more explicitly on racist lines.[62] Then population change is all too easily seen as a security threat to Northern societies. In this interpretation of the global security situation, democracy is a luxury of the rich nations; the rest are doomed to penury and misery in perpetuity.

The consequent policy suggestions from some commentators, like Manfred Worner of NATO, often suggest that threats posed by global population change are external threats to the Northern-dominated status quo, rather than a common challenge to be faced by all humanity. In this case the geopolitical divisions of "them and us" once again posit security as a particular political identity to be defended against external "others" rather than as a common security problem requiring coordinated action and possibly Northern concessions in the face of global inequities.

The Debt Crisis, Structural Adjustment, and Environment

Given these dynamics and the large disparities between North and South, calls to limit global consumption of resources are easily dismissed by those who have relatively few resources to consume. Many delegates from the richer social groups in the South at the 1992 Rio Earth Summit were very insistent that development was the more appropriate international priority than a concern with the environment. The logic of these arguments once again suggests that only development can provide security for Southern populations. But it is precisely the processes of development that have often rendered populations vulnerable to various disasters and hazards by disempowering them, reducing their economic options and making key resources harder to acquire.[63] The point is not that economic change and various forms of empowerment and change are not needed; rather, the question is whether the conventional economic development strategies can provide forms of economic security without undermining environmental sustainability.

A more detailed look at the economic processes of (unsustainable) development in the South clearly reveals that the contemporary structure of the global economy is a very important cause of environmental degradation. Many underdeveloped countries have found their economies constrained by only being able to export a few basic commodities to more wealthy states. Development is constrained by the demands from the rich urban areas of the North for resources which, when extracted from the South, constitute a "shadow economy" of degradation beyond the boundaries of the states in which they are ultimately consumed.[64] Economic diversification is often very difficult in these circumstances. "Development" as advocated by the official agencies charged with its implementation often involves ignoring the specific conditions in particular places in favor of generalized models of modernization.[65] Often these involve improving the infrastructure to export resources, a process which may enrich the local elite, rather than improving the lives of the poorest in the population. These often involve precisely the projects that the elites in the South support to bring in external funding, maintain their power, and attempt to follow Northern routes to economic growth. (This argument also reveals the geopolitical limitations of the Southern critique inherent when overgeneralizing the common interests of elites and populations and suggests the need for precision in dealing with equity concerns.)

Recent literature on environmental change in the Third World documents the damage wrought by many conventional Northern modes of development. The list of destruction and disruption that results is long.[66] Dams flood fertile valleys to provide hydroelectric power to distant urban dwellers. Forests are logged for the benefit of urban-based forestry companies, many of them foreign owned. The habitat destroyed in the process is often the home of indigenous peoples as well as exotic animal and plant species that make up valuable parts of the biological diversity of the planet, play roles in regulating water flows and provide food reserves and medicines for local people.[67] Green revolution technologies may increase the overall productivity of agriculture in specific states, usually those that are wealthy enough to adopt the expensive chemicals, seeds, and equipment required. Or it benefits those more affluent sections of the agricultural population who are involved in the cash economy to a great enough degree to get access to credit and financial assistance.

The marginal farmers, and in particular women, eking out an

existence on the fringes of the agricultural economy, are often forced to move as larger technological farms enclose and consolidate land holdings formerly used for subsistence farming.[68] The agricultural export model of development promoted by Northern institutions in order to provide foreign exchange to pay for international debts has often aggravated this situation. With larger enclosures of land for commercial production the poor move on to marginal lands and steep slopes aggravating soil erosion and fertility loss, which undermines the ecological integrity of large areas. But these patterns of "development" have also turned many poor people into environmental activists struggling to maintain ecosystems intact in the face of external encroachments. Grassroots action, rather than state activity, is very often crucial to environmental protection.[69]

The politics of these changes can be violent and destructive. Conflict over environmental issues is usually caught up with struggles for access to land fought between richer landlords, often with the support of state and military agencies, and poor peasants and farmers whose interests are challenged by the expansion of commercial operations.[70] Structural adjustment policies of the 1980s to deal with the "debt problem" often led to undermining the long-term sustainability of Southern resource bases and made the lot of poor women even more difficult.[71] Within Southern states the maldistribution of wealth and power was often accentuated by these processes. Through the 1980s, largely as a result of the "debt crisis," the net flow of resources internationally was also from poor to rich.[72]

Here once again is the heart of the dilemma that results from linking security to environment. To link these themes requires reformulating at least one of the terms. If sustainable development, which guarantees continued economic growth, is to be secured by new technological management of the globe through new theories of development and intensified resource use, then environmental concerns at the large scale have to be downgraded in importance, if not dismissed altogether. If, on the other hand, security is reformulated to emphasize environmental themes, it will require tackling the belief in the efficacy of technological rationality to solve all problems that obstruct sustainable development. It will also require restructuring the world economy that so profligately uses resources in the North that are gathered at the global scale. Crucially it also suggests that grassroots communities are important to maintaining environments but that in the process they often act in opposition to conventional state-operated development programs.

Geopolitical Security or
Securing Natural Environments

This chapter has suggested the existence of some obvious contradictions if conventional notions of security are uncritically linked to global environmental concerns. Focusing on geopolitical concerns reasserts traditional understandings of foreign policy; primacy is given to military interpretations of options and attention is placed on threats to established political order. When coupled to the access-to-resources theme, fears of demographic competition between North and South, or nuclear weapons and missile proliferation scares, geopolitical specifications of security in terms of external threats once again suggest an American-led security order defined against a variety of external antagonists. The traditional geopolitical notions of security emphasized territorial control, superior alliance systems, and technological superiority in military matters. Insofar as these geopolitical themes dominate discussion and debate about future U.S. global political roles, coupling these themes to environmental security is likely to perpetuate patterns of violence and domination rather than work to protect environments and their peoples.

A further danger of thinking of environment in security terms is that police actions and military interventions may be legitimated. Israel took unilateral action against an Iraqi nuclear reactor in 1981, and the United States objected to a chemical plant in Libya some years later, so precedents for such military intervention exist. The political difficulties of organizing these kinds of operations, given the lack of global consensus on the nature of environmental crises or the culpability of specific societies in environmental degradation, would no doubt be considerable. But one obvious scenario suggests the possibility of a military coalition confronting China over greenhouse and/or CFC emissions if it continues with its coal-based development trajectory and persists with CFC production in violation of international agreements.[73] Might cruise missiles and smart bombs be the solution to future noncompliant rogue CFC plants? While the possibilities of global military action on these issues seems very far-fetched in the mid 1990s, they may become more likely in the future to the extent that environmental degradation is thought of as a traditional security issue.

Of course most advocates of the environmental security approach see its framework as a means to enhance responses to environmental problems and to facilitate diplomatic accommodation prior to the escalation of environmental difficulties to matters of international confrontation. But dangers arise if they fail to adequately take into account how the world is currently constituted by hierarchies of violence and inequality. They also often ignore the fact that those who benefit from the current inequalities have sometimes adopted the logic of environmental security in ways that may act to perpetuate the status quo.[74] If the planet really is in the grave peril that many environmentalists argue it is, then this is no time for political thinking rendered fuzzy by unexamined premises or unacknowledged ethnocentric geopolitical presuppositions.

The discussions of reformulating security also coincide with the widespread recognition of the limits of formal sovereignty in the modern world. Economic globalization, the speed of air travel, the interconnection of many places in the world by phone, fax, and electronic mail, the power of international advertising, and the collective vulnerability of peoples in many places to pollution and environmental degradation all suggest that distance and political boundaries are not as important as they once were. On the other hand the distances between the rich who have access to the airplanes and fax machines and the poor of the planet who do not, seem to many in the South to be getting wider and the chasm more difficult to shout, or scream, across.[75] Where conventional development policies are followed in the South, the gaps between elites connected to the global economy and the rest of the population reliant on local resources also sometimes seem enormous.

Like other projects seeking to rethink politics seriously in light of environmental considerations, the "progressive" environmental security advocates lack a clear political constituency in the North.[76] There is a need to fundamentally rethink international politics, the role of militarization, and the global economy and to challenge the presupposition of a limitless nature that has been taken for granted for so long. It is also necessary to see debt and poverty as an inevitable part of the contemporary political economy, and one that is an important part of contemporary insecurity for much of the world's population. Debt relief and community development are very different security priorities, but ones that are clearly necessary to environmental security understood as the preservation of renewable resources for human survival.

Rethinking global politics along these lines can go still further. Neither the linear logic of realpolitik nor the economic measurements of conventional policy analysis are well suited to tackling the circular interconnections of environmental problems. In Marvin

Soroos's terms, the traditional self-defense focus on security thinking needs to be replaced by a cooperative and preventative approach.[77] Ecological metaphors of context, diversity, interconnection, adaptability, and mutualism might be much more useful understandings of global security than the current physical and mechanical terminology of containment, force, and surveillance. Nature and political entities are not things that can be externally controlled, but must be lived with, cooperated with, and sustainably reproduced. Ideas of resilience, understood in terms of the ability of systems to return to an original state after disruption, may be more appropriate than a formulation of security as applying violence to external threats. In these terms security is not understood as a matter of force, but rather in terms of the construction of community and the practice of mutual cooperation.

These challenges to policy in both the North and the South are profound. The core question in the environmental security debate is whether geopolitical understandings of security will be superseded by new understandings of political community related to the interconnectedness of global ecology. Will security be widely enough understood as a positive policy, and a series of practices taken by actors other than states, to promote biological diversity and sustainable livelihoods, rather than impose conservation in the South, while simultaneously moving to reduce unsustainable economic practices in the North? If global environmental security is understood in these terms then policies of debt relief, grassroots development, democratization, demilitarization, and locally organized resource conservation may follow. In a dramatic change from the state-centric definitions of security, environmental nongovernmental organizations may be important in the provision of environmental security and in new forms of democratic governance.[78] However, a rather obvious cautionary note is clearly in order here: these policies and practices are rarely easy to implement and none offer panaceas in the face of likely opposition.

Discussions of the potential of a global civil society may also be a bit premature in the face of cultural differences and the numerous obstacles to global debate, but recent political activity round the globe has been influenced by transnational social movements, which operate in ways that evade neat encapsulations by state borders.[79] Grassroots movements linked together in numerous networks around the planet are connecting matters of justice and development to matters of cultural survival, human rights, and demilitarization and in the process redefining political activity.[80] In many cases these movements are avoiding the conventional structures of power and trying to open up new ways to act collectively that preserve environments and challenge the conventional assumptions about security, peace, and development in North and South.

In the context of the mid-1990s and the hegemony of liberal internationalism and faith in market-driven social change, the prospects for policies of global environmental security that address questions of justice and new forms of governance may seem somewhat remote. Nonetheless thinking these connections through is important for clarifying the political questions raised by the juxtaposition of security with environment and casting the debate in global terms. If they only serve to clarify the difficulties of ensuring security to much of the world's population then they serve a useful pedagogic purpose. If environmental security can facilitate innovative thinking about the politics of global change then it may yet provide a useful focus for policy proposals. Above all, thinking through these issues makes it clear that a politics concerned about sustainability cannot be simply a repeat of past patterns; new models and modes of analysis are clearly needed whether they are framed in the discourse of security or in some new, yet-to-be-formulated ways. But it is also clear that to think in such new ways will require directly confronting many facets of the political order that geopolitical reasoning has acted to render "secure" for the last half century.

—————————————— NOTES ——————————————

1. Norman Myers, "Environment and Security," *Foreign Policy* no. 47 (1989), pp. 23–41; Jessica Tuchman Mathews, "Redefining Security" *Foreign Affairs* vol. 68, no. 2, 1989, pp. 162–77; and Michael Renner *National Security: the Economic and Environmental Dimensions* (Washington, D.C.: Worldwatch paper 89, 1989).

2. See for example, from a rapidly growing literature, Gro Harlem Brundtland, "The Environment, Security and Development" in *SIPRI Yearbook 1993: World Armaments and Disarmament* (Oxford; Oxford University Press, 1993); Gwyn Prins, ed., *Threats without Enemies: Facing Environmental Inecurity* (London: Earthscan, 1993); Peter Stoett, "The Environmental Enlightenment: Security Analysis Meets Ecology," *Coexistence* vol. 31, no. 2 (1994), pp. 127–47; Eric K. Stern, "Bringing the Environment In: The Case for Comprehensive Security," *Cooperation and Conflict* vol. 30, no. 3 (1995), pp. 211–37; Monica Tennberg, "Risky Business: Defining the Concept of En-

vironmental Security," *Cooperation and Conflict* vol. 33, no. 1, pp. 239–58; Nina Graeger, "Environmental Security," *Journal of Peace Research* vol. 33, no. 1 (1996), pp. 109–16; and Michael Renner, *Fighting for Survival: Environmental Decline, Social Conflict and the New Age of Insecurity* (New York: Norton, 1996).

3. United Nations Development Program, *Human Development Report 1994* (New York: Oxford University Press, 1994).

4. Commission on Global Governance, *Our Global Neighbourhood* (Oxford: Oxford University Press, 1995).

5. On Clinton administration policy, see Geoffrey D. Dabelko and P. J. Simmons, "Environment and Security: Core Ideas and U.S. Government Initiatives," *SAIS Review* vol. 17, no. 1 (1997), pp. 127–46.

6. Marc A. Levy, "Is the Environment a National Security Issue?" *International Security* vol. 20, no. 2 (1995), pp. 35–62. For a particularly lucid classification of the interconnections between the environment and security, see Lothar Brock, "Peace through Parks: The Environment on the Peace Research Agenda," *Journal of Peace Research* vol. 28, no. 4 (1991), pp. 407–24.

7. Thomas Homer-Dixon, "On the Threshold: Environmental Changes as Causes of Acute Conflict," *International Security* vol. 16, no. 1 (1991), pp. 76–116; Thomas Homer-Dixon, "Environmental Scarcities and Violent Conflict: Evidence from Cases," *International Security* vol. 19, no. 1 (1994), pp. 5–40.

8. David D. Dabelko and Geoffrey D. Dabelko, "Environmental Security: Issues of Conflict and Redefinition," *Environment and Security* vol. 1, no. 1 (1996) pp. 23–49.

9. David D. Dabelko and Geoffrey D. Dabelko, "The International Environment and the U.S. Intelligence Community," *Intelligence and Counterintelligence* vol. 6, no. 1 (1993), pp. 21–41; Simon Dalby, "Security, Intelligence, The National Interest and the Global Environment," *Intelligence and National Security* vol. 10, no. 4 (1995), pp. 175–97.

10. For the classic statement on security and ethnocentrism, see Ken Booth, *Strategy and Ethnocentrism* (London: Croom Helm, 1979). On ethnocentrism more generally, see Edward Said, *Orientalism* (New York: Vintage, 1979); and Robert Young, *White Mythologies: Writing History and the West* (London: Routledge, 1990).

11. David Campbell, *Writing Security: American Foreign Policy and the Politics of Identity* (Minneapolis: University of Minnesota Press, 1992); Cynthia Weber, *Simulating Sovereignty: Intervention, the State and Symbolic Exchange* (Cambridge, U.K.: Cambridge University Press, 1995).

12. Daniel Deudney, "The Mirage of Ecowar: The Weak Relationship among Global Environmental Change, National Security and Interstate Violence" in I. H. Rowlands and M. Greene, eds., *Global Environmental Change and International Relations* (London: Macmillan, 1992); Lothar Brock, "Security through Defending the Environment: An Illusion?" in E. Boulding, ed., *New Agendas for Peace Research: Conflict and Security Reexamined* (Boulder, Colo.: Lynne Rienner, 1992); and Jyrki Kakonen, ed., *Green Security or Militarized Environment* (Aldershot: Dartmouth, 1994).

13. John Agnew and Stuart Corbridge, *Mastering Space: Hegemony, Territory and International Political Economy* (New York: Routledge, 1995); Michael Redclift, *Wasted: Counting the Costs of Global Consumption* (London: Earthscan, 1996).

14. Simon Dalby, "Contesting an Essential Concept: Reading the Dilemmas in Contemporary Security Discourse," in Keith Krause and Michael Williams, ed., *Critical Security Studies: Concepts and Cases* (Minneapolis: University of Minnesota Press, 1997), pp. 3–31.

15. United Nations, *World Development Report 1994*, p. 35.

16. Geoff Tansey, Kath Tansey, and Paul Rogers, eds., *A World Divided: Militarism and Development after the Cold War* (London: Earthscan, 1994).

17. Andrew Hurrell and Ngaire Woods, "Globalization and Inequality," *Millennium* vol. 24, no. 3 (1995), pp. 447–70.

18. Bradley S. Klein, *Strategic Studies and World Order: The Global Politics of Deterrence* (Cambridge, U.K.: Cambridge University Press, 1994).

19. It is worth emphasizing that access to resources is only part of the larger pattern of international political relations during the Cold War. For discussion, see Ronnie D. Lipschutz and John P. Holdren, "Crossing Borders: Resource Flows, the Global Environment, and International Security," *Bulletin of Peace Proposals* vol. 21, no. 2 (1990), pp. 121–33.

20. Richard J. Barnett and John Cavanagh, *Global Dreams: Imperial Corporations and the New World Order* (New York: Simon and Schuster, 1994).

21. Alvaro Soto, "The Global Environment: A Southern Perspective," *International Journal* vol. 47, no. 4 (1992), pp. 679–705; N. Middleton, P. O'Keefe and S. Moyo, *Tears of the Crocodile: From Rio to Reality in the Developing World* (London: Pluto, 1993).

22. This theme runs through contemporary discussions of security. See Simon Dalby, "Canadian National Security and Global Environmnetal Change," in Jim Hansson and Susan McNish eds., *"Canadian Strategic Forecast 1997—Canada and the World: Non-Traditional Security Threats* (To-

ronto: Canadian Institute for Strategic Studies, 1997), pp. 19–40; and Ronnie Lipschutz, ed., *On Security* (New York: Columbia University Press, 1995).

23. Simon Dalby, *Creating the Second Cold War: The Discourse of Politics* (London: Pinter, and New York: Guilford, 1990).

24. R. B. J. Walker, *Inside / Outside: International Relations as Political Theory* (Cambridge, U.K.: Cambridge University Press, 1993).

25. Daniel Deudney. "The Mirage of Ecowar," and Lothar Brock, "Security through Defending the Environment," op.cit. note 12. For local contexts and what is being sustained, see Dharam Ghai, ed., *Development and Environment: Sustaining People and Nature* (Oxford: Blackwell, 1994).

26. J. J. Romm, *Defining National Security: The Non-Military Aspects* (New York: Council on Foreign Relations, 1993); Michael T. Klare and Daniel C. Thomas, eds., *World Security: Challenges for a New Century* (New York: St. Martins, 1994).

27. See David Campbell, *Politics without Principle: Sovereignty, Ethics, and the Narratives of the Gulf War* (Boulder, Colo.: Lynne Rienner, 1993); and John Agnew, *Geopolitics: Revisioning World Politics* (London: Routledge, 1998).

28. Robert D. Kaplan, "The Coming Anarchy," *The Atlantic Monthly* vol. 273, no. 2 (1994), pp. 44–76. For a detailed critique, see Simon Dalby, "The Environment as Geopolitical Threat: Reading Robert Kaplan's Coming Anarchy," *Ecumene* vol. 3, no. 4 (1996), pp. 472–96.

29. Wolfgang Sachs, "Global Ecology and the Shadow of Development," in Wolfgang Sachs, ed., *Global Ecology: A New Arena of Political Conflict* (London: Zed, 1993).

30. Manfred Worner, "Global Security: The Challenge to NATO," in E. Grove, ed., *Global Security: North American, European and Japanese Interdependence in the 1990s* (London: Brassey's, 1991), p. 102.

31. Worner, "Global Security," pp. 103–104.

32. Worner,"Global Security," pp. 104–105.

33. For analysis of the more general question of national security in the context of economic interdependence, see Beverley Crawford, "The New Security Dilemma under International Economic Independence," *Millennium* vol. 23, no. 1 (1994), pp. 25–55.

34. But it is also worth noting that these arguments have often been used either as a substitute, or as a support, for policies initiated for other political reasons. See Ronnie D. Lipschutz, *When Nations Clash: Raw Materials, Ideology and Foreign Policy* (Cambridge, Mass.: Ballinger, 1989).

35. Amory Lovins and L. Hunter Lovins, *Brittle Power: Energy Strategy for National Security* (Andover, Mass.: Brick House, 1982).

36. B. R. Inman, J. S. Nye, W. J. Perry, and R. K. Smith, "Lessons from the Gulf War," *The Washington Quarterly* vol. 15, no. 1 (1992), pp. 67–8.

37. Sten Nilsson and David Pitt, *Protecting the Atmosphere: The Climate Change Convention and Its Context* (London: Earthscan, 1994).

38. Anil Agarwal and Sunita Narain, *Global Warming in an Unequal World: A Case of Environmental Colonialism* (New Delhi: Centre for Science and Environment, 1991). The essential point of this argument is captured in the Scott Willis political cartoon on the front cover of this report. It shows a diminuitive peasant about to chop down a tree. However, just prior to the first blow of the axe, a large male figure standing in an enormous automobile, with license plate reading "developed countries," and with actively emitting exhaust pipe prominently displayed, remonstrates with the peasant, saying "Yo! Amigo!! you can't cut that tree. We need it to stop the greenhouse effect."

39. Vandana Shiva, "Conflicts of Global Ecology: Environmental Activism in a Period of Global Reach," *Alternatives* vol. 19, no. 2 (1994), pp. 195–207.

40. Vandana Shiva, "The Greening of Global Reach," in Wolfgang Sachs, ed., *Global Ecology: A New Arena of Political Conflict* (London: Zed, 1993), pp. 151–52.

41. Richard Benedick, *Ozone Diplomacy: New Directions in Safeguarding the Planet* (Cambridge, Mass.: Harvard University Press, 1991).

42. Elinor Ostrom, *Governing the Commons: The Evolution of Institutions for Collective Action* (Cambridge, U.K.: Cambridge University Press, 1990).

43. Rachael M. McCleary, "The International Community's Claim to Rights in Brazilian Amazonia," *Political Studies* vol. 39, no. 4 (1991), pp. 691–707.

44. On this theme, see Andrew Hurrell, "A Crisis of Ecological Viability? Global Environmental Change and the Nation State," *Political Studies* vol. 42, (1994), pp. 146–65; and Ken Conca, "Rethinking the Ecology-Sovereignty Debate," *Millennium* vol. 23, no. 3 (1994), pp. 701–11.

45. Jose Goldemberg and Eunice Ribeiro Durham, "Amazonia and National Sovereignty," *International Environmental Affairs* vol. 2, no. 1 (1990), pp. 22–39; Terry Terriff, "The 'Earth Summit': Are There Any Security Implications?" *Arms Control* vol. 13, no. 2 (1992), pp. 163–90; and A. Pinheiro and P. Cesar, "A Vision of the Brazilian National Security Policy on the Amazon," *Low Intensity Conflict and Law Enforcement* vol. 3 (1994), pp. 387–409.

46. R. P. Guimaraes, *The Ecopolitics of Development in the Third World: Politics and Environment in Brazil* (Boulder, Colo.: Lynne Rienner, 1991); Suzanna Hecht and Alexander Cockburn, *The Fate of the Forest: Developers, Destroyers and Defenders of the Amazon* (Harmondsworth: Penguin, 1990).

47. Bruce Albert, "Indian Lands, Environmental Policy and Military Geopolitics in the Development of the Brazilian Amazon: The Case of the Yanomami," *Development and Change* vol. 23, no. 1 (1992), pp. 35–70.

48. Franz Broswimmer, "Botanical Imperialism: The Stewardship of Plant Genetic Resources in the Third World," *Critical Sociology* vol. 18, no. 1 (1991), pp. 3–17.

49. The Centre for Science and Environment, "Statement on Global Environmental Democracy," *Alternatives* vol. 17, no. 2 (1992), pp. 261–79; and Tom Athanasiou, *Divided Planet: The Ecology of Rich and Poor* (Boston: Little Brown, 1996).

50. Marian A. L. Miller, *The Third World in Global Environmental Politics* (Boulder, Colo.: Lynne Rienner, 1995).

51. Thanks to Daniel Deudney for helpful suggestions on this point.

52. Kent Butts, "Why the Military Is Good for the Environment," in Jyrki Kakonen, ed., *Green Security or Militarized Environment* (Aldershot: Dartmouth, 1994).

53. Mohammad Ayoob, *The Third World Security Predicament* (Boulder, Colo.: Lynne Rienner, 1995); Brian Job, ed., *The Insecurity Dilemma* (Boulder, Colo.: Lynne Rienner, 1992).

54. Nancy L. Peluso, "Coercing Conservation: The Politics of State Resource Control," in Ronnie D. Lipschutz and Ken Conca, eds., *The State and Social Power in Global Environmental Politics* (New York: Columbia University Press, 1993).

55. Bruce Byers, "Armed Forces and the Conservation of Biological Diversity," in Jyrki Kakonen, ed., *Green Security or Militarized Environment* (Aldershot: Dartmouth, 1994).

56. J. Mackenzie, *The Empire of Nature: Hunting, Conservation and British Imperialism* (Manchester: Manchester University Press, 1989).

57. On "desecuritization" see Ole Waever, "Securitization and Desecuritization," in Ronnie D. Lipschutz, ed., *On Security* (New York: Columbia University Press, 1995). Versions of this argument are made also by both Daniel Deudney in "The Mirage of Ecowar" and Lothar Brock in "Security through Defending the Environment," op.cit. note 12.

58. Norman Myers, *Ultimate Security: The Environmental Basis of Politi-*

cal Stability, (New York: Norton, 1993). Also see George D. Moffett, *Critical Masses: The Global Population Challenge* (New York: Viking, 1994).

59. Dennis Pirages, "Demographic Change and Ecological Insecurity," in Michael T. Klare and Daniel C. Thomas, eds., *World Security: Challenges for a New Century* (New York: St. Martins, 1994).

60. Matthew Connelly and Paul Kennedy, "Must It Be the Rest Against the Rest?" *The Atlantic Monthly* vol. 274, no. 6 (1994), pp. 61–83; and Paul Kennedy, *Preparing for the Twenty-First Century* (New York: HarperCollins, 1993).

61. Ole Waever, Barry Buzan, Morten Kelstrup, and Pierre Lemaitre, *Identity, Migration and the New Security Agenda in Europe* (London: Pinter, 1993).

62. Nicholas Eberstadt, "Population Change and National Security," *Foreign Affairs* vol. 70, no. 3 (1991), pp. 115–31; Anthony H. Richmond, *Global Apartheid: Refugees, Racism, and the New World Order* (Toronto: Oxford University Press, 1994).

63. Piers Blakie, Terry Cannon, Ian Davis, and Ben Wisner, *At Risk: Natural Hazards, People's Vulnerability and Disasters* (London: Routledge, 1994).

64. For discusion of "shadow economies," see Jim MacNeill, Peter Winsemius, and Taizo Yakushiji, *Beyond Interdependence: The Meshing of the World's Economy and the Earth's Ecology* (Oxford: Oxford University News, 1991).

65. Jonathan Crush, ed., *Power of Development* (London: Routledge, 1995).

66. Bruce Rich, *Mortgaging the Earth: The World Bank, Environmental Impoverishment and the Crisis of Development* (Boston: Beacon Press, 1994).

67. Vandana Shiva, *Staying Alive: Women, Ecology and Development* (London: Zed, 1988); Madhav Gadgil and Ramachandra Guha, *Ecology and Equity: The Use and Abuse of Nature in Contemporary India* (London: Routledge, 1995).

68. Barbara R. Johnston, ed., *Who Pays the Price? The Sociocultural Context of Environmental Crisis* (Washington, D.C.: Island Press, 1994).

69. Robin Broad, "The Poor and the Environment: Friends and Foes," *World Development* vol. 22, no. 6 (1994), pp. 811–22; Alan Durning, *Action at the Grassroots: Fighting Poverty and Environmental Decline* (Washington, D.C.: Worldwatch Institute Paper 88, 1989).

70. Daniel Faber, *Environment Under Fire* (New York: Monthly Review Press, 1993).

71. Dianne Rocheleau, Barbara Thomas-Slayter, and Esther Wangari, eds., *Feminist Political Ecology: Global Issues and Local Experiences* (New York: Routledge, 1996).

72. Susan George, *The Debt Boomerang: How Third World Debt Harms Us All* (Boulder, Colo.: Westview, 1992).

73. Daniel Deudney, "The Mirage of Ecowar," op.cit. note 13.

74. For an argument that the 1992 Earth Summit has effectively put this security option into action, see Pratap Chatterjee and Mathias Finger, *The Earth Brokers: Power, Politics and World Development* (New York: Routledge, 1994).

75. See the discussion on this theme in "Against Global Apartheid," Special Issue of *Alternatives* vol. 19, no. 2 (1994). Titus Alexander uses similar language and reasoning in his *Unravelling Global Apartheid* (Cambridge, Mass.: Polity, 1996).

76. John McCormick, *The Global Environmental Movement* (New York: Wiley, 1995); Andrew Dobson, *Green Political Thought* (London: Routledge, 1995).

77. Marvin S. Soroos, *The Endangered Atmosphere: Preserving a Global Commons* (Columbia: University of South Carolina Press, 1997); Arran E. Gare, *Postmodernism and the Environmental Crisis* (London: Routledge, 1995); and Gwyn Prins, "Politics and the Environment," *International Affairs* vol. 66, no. 4 (1990), pp. 711–30.

78. R. B. J. Walker, *One World, Many Worlds: Struggles for a Just World Peace* (Boulder, Colo.: Lynne Rienner, 1988); Fen Osler Hampson and Judith Reppy, eds., *Earthly Goods: Environmental Change and Social Justice* (Ithaca, N.Y.: Cornell University Press, 1966).

79. Ronnie D. Lipschutz, *Global Civil Society and Global Environmental Governance: The Politics of Nature from Place to Planet* (Albany: State University of New York Press, 1996); Paul Wapner, *Environmental Activism and World Civic Politics* (Albany: State University of New York Press, 1996); and Richard Peet and Michael Watts, eds., *Liberation Ecologies: Ecologies, Environment, Development, Social Movements* (New York: Routledge, 1996).

80. R. B. J. Walker, "Social Movements/World Politics," *Millennium*, vol. 23, no. 3 (1994), pp. 669–700.

8

Environmental Security

A Critique

Daniel H. Deudney

Like all biological organisms, humans are bound by bodily needs to an inescapable dependence on their physical environment. In the industrial era of the last two centuries the explosive progress in science and technology and the emergence of societies of unprecedented wealth seemed to have loosened the iron grip of natural scarcity upon human life, presaging what Karl Marx and other nineteenth-century visionaries of permanent abundance termed the end of prehistory and the beginning of human history. But over the last several decades this heady vision of an escape from nature has proven to be a fantastic illusion. Environmentally abusive practices, compounded by the sheer weight of human numbers and the power of industrial technology, have increasingly degraded a wide range of vital and interconnected resources that humans had been able to take for granted throughout history.[1] Over the next several decades human population is likely to nearby double again, and economic output increase three- to five-fold, setting the stage for either the collapse of industrial civilization or a far-reaching "green" transformation of all aspects of human life. Far from ending, history promises to return to an era in which humanity's relationship with nature again will be central and precarious. In the face of these ominous developments, the quest for a new political paradigm to conceptualize problems and motivate action has immense practical importance. But what should this paradigm be?

The "Environmental Security" Movement

Because the institutions and ideologies of nation-state and interstate conflict are so hegemonic in both world politics and international relations theory, and because violence has historically occurred over competition for environmental scarcities there is a natural tendency for people to think about environmental problems in terms of national security. Initial moves to connect security and national security with environmental issues were made by Lester Brown of the Worldwatch Institute in 1977.[2] More broadly, Richard Ullman proposed "redefining security" to encompass a wide array of threats, ranging from earthquakes to environmental degradation.[3] Arthur Westing pointed to the destruction of the environment caused by war and hypothesized that interstate war and other forms of violence would result from resource scarcity and environmental degradation.[4] Patricia Mische proposed to "reconceptualize sovereignty" in order to focus on "ecological security."[5] Most of the pioneering conceptual work on environmental security was done by advocates of greater environmental awareness. Such concepts were advanced to prevent an excessive focus on military threats during the renewed Cold War tensions and heightened "national security" concerns of the late 1970s and early 1980s. They also were extrapolations from the fears of resource wars that had been widely discussed in the wake of the oil crises of 1973 and 1979, the formation of commodity cartels, and rapid price rises of oil and other earth minerals during the 1970s.

By the late 1980s and early 1990s environmental security became a broad movement, had generated an empirical research agenda, and had begun to shape policy. Numerous environmental advocates, including (but not limited to) Michael Renner, Jessica Tuchman Mathews, Norman Myers, and Gwyn Prins, wrote in favor of redefining national security to encompass resource and environmental threats.[6] Due to the interest and support of several major foundations in the United States, numerous conferences were held, and large numbers of researchers began addressing issues of environmental security. A major research effort, headed by Thomas Homer-Dixon of the University of Toronto and partially sponsored by the American Academy of the Arts and Sciences, has explored more systematically the links between environmental degradation, renewable resource scarcity, and violent conflict.

As the Cold War waned in the the late 1980s, environmental security began to attract the interest and support of many associated with military organizations who saw environmental missions as a means to maintain financial support and organizational significance. Others saw environmental deterioration, particularly in Third World countries, as part of an ominous new threat to Western interests and world order. This new security fear was catalyzed by Robert Kaplan's horrific travelogue, "The Coming Anarchy," and his widely cited conclusion that "the environment is *the* national security issue of the 21st century."[7] With apocalyptic speculations of about chaos in the Third World, environmental security became a contender in the United States' effort to formulate a new post–Cold War foreign policy.[8] The U.S. Congress established and funded several environmental security initiatives to begin addressing military sources of environmental degradation, and the U.S. Department of Defense launched several environmental security initiatives and organizational changes.[9]

Initially the environmental security paradigm and agenda seemed straightforward and noncontroversial. But in the early 1990s, a range of objections and doubts were raised by this author,[10] Mattias Finger, Simon Dalby, Ken Conca, and Marc Levy,[11] all of whom are strongly sympathetic to environmental concerns. Since these debates were first joined, extensive research has been undertaken, and heated debates about environmental security have occurred in many policy and academic fora.[12] This chapter revisits, refines, and extends the arguments against the environmental security paradigm and program.

Overall, skepticism is not only still warranted, but confirmed and strengthened. Specifically, I make three claims. First, it is analytically misleading to think of environmental degradation as a national security threat because the traditional focus of national security— interstate violence—has little in common with either environmental problems or solutions. Second, the effort to harness the emotive power of nationalism to help mobilize environmental awareness and action may prove counterproductive by undermining globalist political sensibility. And third, environmental degradation is not very likely to cause interstate wars.

I. Assessing Similarities and Differences

The first major strand of environmental security thinking has sought to redefine national security or, more broadly, security, to encompass threats to societal welfare that have traditionally been outside their domain. Historically, such conceptual shifts have often accompanied important changes in politics, as new phrases are coined and old terms are appropriated for new purposes.[13] Epochal developments like the emergence of capitalism, the growth of democracy, and the end of slavery were accompanied by shifting, borrowing, and expanding political language. The wide-ranging contemporary conceptual ferment in the language used to understand and act upon environmental problems is therefore not either an unexpected or an intrinsically undesirable development.

But not all neologisms and linkages are equally plausible or useful. Until this recent movement of reconceptualization, the concept of national security (as opposed to national interest or welfare) has been centered upon *organized violence*.[14] As many political theorists have insisted, security from violence is a fundamental human need because loss of life prevents the enjoyment of all other goods. Although resource factors traditionally were understood as contributing to state capacities to wage war and achieve security from violence, they were security issues because of their links to state warmaking capability rather than intrinsically seen as security threats in their own right.

Before either expanding the concept of national security to encompass both environmental and violence threats, or redefining national security or security to refer mainly to environmental threats, it is worth examining just how much the national pursuit of security from violence has in common with environmental problems and their solutions.

Military Sources of Environmental Degradation

Military violence and environmental degradation are linked directly in at least three major ways. First, the pursuit of national security from violence through military means consumes resources (fiscal, organizational, and leadership) that could be spent on environmental restoration. Because approximately one trillion dollars is spent world-wide on military activities, substantial resources are involved. However, this relationship is not unique to environmental concerns, and there is no guarantee that money saved from military cut-backs would be spent on environmental restoration. And the world can afford environmental restoration without cutting military expenditures.

Second, war is directly destructive of the environment. Some of this destruction is an unintentional effect of war, while some of it is the intentional destruction of the natural environment, or environmental warfare. In ancient times, the military destruction of olive groves in Mediterranean lands contributed to the long-lasting destruction of the land's carrying capacities[15] (although military destruction was probably far less significant than mundane agricultural, forestry, and grazing practices). More recently, the United States' bombardment and use of defoliants in Indochina caused significant environmental damage.[16] During the Gulf War of 1991, Saddam Hussein's retreating forces set the oil fields of Kuwait on fire, causing massive ecological damage and resource wastage.[17] Most ominously, extensive use of nuclear weapons would have significant impacts on the global environment, including altered weather patterns (i.e., "nuclear winter") and further depletion of the ozone layer.[18] Even a "conventional" war in highly industrialized areas could be environmentally catastrophic due to the release of radiation from civilian nuclear power plants. Awareness of the potential environmental effects of nuclear weapons played an important role in mobilizing popular resistance to the arms race and in generally delegitimizing the use of nuclear explosives as weapons, first in the concern over atmospheric testing during the late 1950s and early 1960s and then over nuclear winter in the mid 1980s.

Third, preparation for war poses a significant environmental burden. Preparation for modern industrial and nuclear war consumes large qualtities of metal and fuel, and has generated large quantities of toxic and radioactive waste. Over the last half-century, the nuclear weapons states, particularly the United States and the Soviet Union, generated enormous quantities of radioactive waste as a by-product of the nuclear weapons development and production.[19] Safely handling these materials is expensive and requires highly competent organizations. And as with other large and complex technical systems, accidents, catastrophic on a local and sometimes regional scale, are inevitable. For example, in 1957 the waste dump at a Soviet nuclear weapons fabrication facility exploded and burned, spreading radioactive materials over a large area near the Urals.[20]

In summary, war and the preparation for war are clearly environmental problems and consume resources that could be used to ameliorate environmental degradation. These environmental impacts mean that the dependence of the international system upon the use and threat of large-scale violence to resolve conflicts has costs beyond the intentional loss of life and destruction. Nevertheless, most environmental degradation is not caused by war and preparation for war. Even if the direct environmental effects of preparing for and waging war were completely eliminated, most environmental degradation would remain. Environmental degradation's main sources and solutions are found outside the domain of the traditional national security system related to violence.

Comparing Environmental and Military Security Issues

The war system is a significant but limited source of environmental destruction, but in what ways is environmental degradation a threat to national security? One answer to this question, advanced by many analysts of environmental security, is to broaden the definition of national security to encompass environmental harms. Making such an identification can be useful conceptually and analytically if the two phenomena—security from violence and from environmental threats—are similar. How similar are they with regard to the *type* of threat, the *source* of threat, degree of *intentionality*, and the types of *organizations* involved? (See table 8.1.)

First, environmental degradation and interstate violence both entail threats to life and property, but they are very different in character. Both violence and environmental degradation may kill people and may reduce human well-being, but not all threats to life and property are threats to security. Disease, aging, crime, and accidents routinely destroy life and property, but we do not think of them as national security threats or even threats to security. (Crime is a partial exception, but crime is a security threat at the individual level because crime involves violence.) And when an earthquake or hurricane strikes with great force, we speak of "natural disasters," or designate "national disaster areas," but we do not speak about such events threatening national security. If everything that causes a reduction in human well-being is labeled a security threat, the term looses any analytical usefulness and becomes a loose synonym of "bad."

Second, the scope and source of threats to environmental well-being and national-security-from-violence are very different. There

Table 8.1 Conceptual and organizational dissimilarities

	TRADITIONAL NATIONAL SECURITY	GLOBAL HABITABILITY
Type of Threat	Violent death Destruction of property Loss of political independence	Wide range of harms: aesthetic, disease, ecological integrity, and resource degradation
Source of Threat	Mainly external Mainly other states armed with weapons	Both internal and external Wide range of sources: individuals, corporations and governments
Degree of Intentionality	Mainly direct and high	Mainly unintentional side-effects of routine activities
Types of Organizations Involved	Specialized Secretive Removed from civil society	All sizes and types In-situ changes in many mundane activities: land use, waste treatment, agriculture, industrial processes, building design, and transportation systems

is nothing about the problem of environmental degradation that is particularly "national" in character. Few environmental problems afflict just one nation-state because they often spill across international borders, or affect the global commons beyond state jurisdiction. At the same time, most environmental problems are not "international," because many perpetrators and victims are within the borders of one nation-state. Individuals, families, communities, other species, and future generations are harmed. A collapse of the biosphere would surely destroy nation-states as well as everything else, but there is nothing distinctively national about either the causes, the harms, or the solutions that warrants privileging the national grouping.

A third dissimilarity between environmental well-being and national-security-from-violence stems from the differing degrees of *intention* involved. Interstate violence typically involves a high degree of intentional behavior. Organizations are mobilized, weapons procured, and wars waged with relatively definite aims in mind. In contrast, environmental degradation is largely the unintentional side-effect of many other activities. With the exception of "environmental warfare" people rarely act with the intentional goal of harming the environment.

Fourth, organizations that provide protection from violence differ greatly from those in environmental protection. National-security-from-violence is conventionally pursued by organizations with three distinctive features. Military organizations are secretive, extremely hierarchical, and centralized, and typically deploy expensive, highly specialized, and advanced technologies. Also, citizens typically delegate the goal of achieving national security to remote and highly specialized organizations that are far removed from civil society. Lastly, the specialized professionals who staff national security organizations are trained in the arts of killing and destroying.

In contrast, responding to the environmental problem requires very different approaches and organizations. Change is required in aspects of virtually all mundane activities from house construction, farming techniques, sewage treatment, factory design, and land use. The routine behavior of virtually everyone must be altered. This requires behavior modification *in situ*. And the professional ethos of environmental restoration is husbandmanship—more respectful cultivation and protection of plants, animals, and the land.

In summary, national-security-from-violence and environmental habitability are far more dissimilar than similar. Given these differences, linking them via redefinition risks creating a conceptual muddle rather than a paradigm or worldview shift. If all the forces and events that threaten life, property, and well-being (on a large scale) are understood as threats to national security, the term will come to be drained of useful meaning. This is even more of a problem for "comprehensive security" paradigms that aggregate all threats to all subjects. If all large-scale evils become threats to security, the result will be a *dedefinition* rather than a *redefinition* of security. To speak meaningfully about actual distinct and different problems it will be necessary to invent new or redefine old words to serve the role performed by the old spoiled ones.

II. "Environmental Security" as a Motivational Strategy

Despite the great differences between the issues of traditional national security and the environment, some environmentalists want to conflate or link them as part of a motivational strategy. Because people take national security threats seriously they are willing to bear heavy costs, in financial and material resources, in organizational commitment and attention, and in compromised liberties and lost lives. Advocates of environmental security reason that if people reacted as urgently and effectively to environmental problems as to the national-security-from-violence problem, then much more effort and resources would be directed to environmental problems. Thus, one aim of redefined national security or environmental security is not to describe or understand the world more accurately, but rather to stimulate, motivate, and inspire action. Environmental security is thus part of a rhetorical and psychological strategy to redirect social energies now devoted to war and interstate violence toward environmental amelioration. In keeping with this purpose, much of the environmental security literature aims to persuade and inspire as much as to analyze and inform.

The Moral Equivalent of War

This motivational strategy is neither original nor unique to the environmental cause. The effort to establish what William James called a "moral equivalent to war" has long been the goal of social reformers.[21] But channeling the energies behind war into constructive directions by this kind of rhetorical linkage has not been particularly successful. In the United States this kind of strategy has been widely tried, as political leaders and their speech writers have launched a "War on Poverty," a "War on Crime," and a "War on Drugs." But these social problems—like the environment—have little in common with the pursuit of national security from violence, and they have proven deeply intractable. As Ken Conca and others have observed, the discourse of national security has a set of powerful associations that cannot simply be redirected.[22]

An even more serious problem with this motivational strategy is that it might have adverse side-effects to the extent that it is effective. Enhanced concern for the environment because it is perceived as a security problem might alter environmental politics in very negative ways. National security claims are politically potent because they have been connected to state institutions, national identities, and international war. It is improbable that these potent forces can be tapped for environmental efforts without bringing into environmental politics the conflictual, parochial, and zero-sum assumptions, norms, practices, and institutions that currently predominate in the domain of national security. In short, there is a danger, as Jyrki Kakonen puts it, of inadvertantly producing a "militarized environment" rather than "green security."[23]

To assess the possible unintended side-effects of "environmental security" as a motivational strategy, it is necessary to begin by recalling the powerful trinity of the state, nation, and war that largely defines national security in contemporary world politics. State institutions, national identities, and interstate war have such salience and persistence in world politics because they powerfully reinforce one another. As the historian Michael Howard has observed: "Self-consciousness as a Nation implies, by definition, a sense of differentiation from other communities, and the most memorable incidents in the group memory usually are of conflict with, and triumph over, other communities. It is in fact very difficult to create national self-consciousness *without* a war."[24] States enter the equation by making war, which in turn strengthens states. Summarizing the dominant view of political scientists and historians, Charles Tilly observes that "states make war and war makes states."[25] States build and sustain national political identities in the populations within their borders through educational systems, public ceremonies, and direct indoctrination that employ memories of war. This dominant approach to framing and pursuing national security originated in early modern Europe, has subsequently become global in scope, and has proven deeply resistant to fundamental reform.

To assess further the potential risks of the environmental security motivational strategy, it is instructive to compare the specific interrelated assumptions, norms, ideologies, identities, and institutions associated with the national war state and those associated with environmental sustainability (see table 8.2).

At first glance, the most attractive feature of linking fears about environmental threats with national security mentalities is the sense of urgency created, and the corresponding willingness to accept great sacrifice. If the basic habitability of the planet is being undermined, then some crisis mentality is warranted. But it is difficult to engender a sense of urgency and a willingness to sacrifice for extended periods of time. Crises call for resolution, and the patience of a mobilized populace is rarely long. For most people exertion in a crisis is motivated by a desire to return to normalcy, for the problem to be resolved once and for all. Wars demand victory and a return to peace. Such a cycle of passivity and arousal is not likely to make much of a contribution to establishing enduring patterns of environmentally sound behavior.

Table 8.2 Traditional national security vs. global sustainability paradigms

	CONVENTIONAL NATIONAL SECURITY	CONSEQUENCE FOR ENVIRONMENTAL POLITICS	GLOBAL SUSTAINABILITY PARADIGM
Institution	Strong state	War against society Coercive conservation Eco-totalitarianism	Strengthen local and global
Identity	Nationality	Reinforce "us" vs. "them"	Bioregional, global, planetary
Authority	State sovereignty	Impede interstate agreements	Diffuse & divided; future generations and other species

Furthermore, "crash" solutions are often bad ones—more expensive, centralized, and poorly designed than other collective efforts.

Framing the environmental problem as a national security threat is also likely to entail the expansion of state capabilities to regulate and manage the environment. In many developing countries, states are simultaneously weak and oppressive and political identities are rarely national. In this context, framing environmental problems as national security threats could provide a mandate for coercion. In Africa where the pattern of political identities does not correspond to the borders of the state system, states tend to be both fragile and despotic. Already many states in the developing world are practicing what Nancy Peluso has termed "coercive conservation," the use of state power to dispossess locals of their traditional natural resources in order to benefit state elites and multinational corporations.[26] And as the anthropologist Jason Clay has pointed out, in many parts of the developing world, and particularly its tropical rain forests, states are dispossessing and destroying indigenous peoples whose claim to distinct national identity is high but whose potential for statehood is very low or nonexistent.[27]

Expanding the security state also puts individual freedom at risk. As James Der Derian has pointed out, "to secure" also means "to tie down" or "prevent from moving."[28] A "security jacket" protects by confining. State actions to secure against a threat often involve erosion of individual liberty and greater restraint and oppression. Strong states have been the greatest source of security threats in the twentieth century: authoritarian and totalitarian states have murdered more of their citizens than died in all interstate wars, and most deaths in war occurred in conflicts started by aggressive authoritarian and totalitarian states.[29] Given this record, Eric Stern's agenda of "comprehensive security" and Norman Myers's agenda of "ultimate security" have a sinister potential.[30] Because almost all human activities affect the environment in some way or another, assigning states the task of environmental security could provide the foundation for an eco-totalitarianism. In short, a state providing comprehensive security will also assume total control.

Contemporary national security is also closely connected to the institution of state sovereignty. In the international "society of states," sovereignty has come to mean the existence in a polity of a final and undivided authority over a particular territory, which only states can possess, and the reciprocal recognition of this authority which states extend to one another.[31] This system of legitimate authority has marginalized the claims to sovereignty, autonomy, and authority advanced by other actors both within states and outside them. When issues of national security are at stake, states tend to be highly jealous of their sovereign prerogatives because they fear that its loss or dilution will leave them subordinate and therefore vulnerable. But responding to international and global environmental problems often requires arrangements that divide and pool authority, and thus run against the normal practices of state sovereignty. Therefore, an enhanced state concern for its sovereign prerogatives could greatly impede international environmental cooperation.

Several of these patterns can be seen in the energy security crisis in the United States during the 1970s. Aiming for the "moral equivalent of war" President Carter sought to mobilize public awareness and forge a public consensus about energy conservation, but once the immediate symptoms of the problem receded, public concern subsided. The "energy independence" initiatives advanced by both parties in Congress and the White House emphasized nuclear energy and expanded fossil fuel production, and largely neglected less-centralized, less-capital intensive, and less-environmentally destructive alternatives. In order to insure that the heavily subsidized macroenergy projects would not be impeded by citizen participation and environmental reviews, Carter proposed an "energy mobilization board" with broad powers to override environmental restraints. And a Byzantine system of price controls was imposed to supplant market mechanisms.[32]

Another apparent fit between the "national security" mentality and the environmental problems is the tendency to use worst-case scenarios as the basis for planning. However, military organizations are not unique in this regard. The insurance industry routinely prepares for the worst possibilities, and many fields of engineering, such as aeronautical design and nuclear power plant regulation, also employ extremely conservative planning assumptions. Therefore, it is not necessary for environmental policy to be modeled after national security and military organizations to achieve risk-averse planning.

Third, the conventional national security mentality and its organizations are deeply committed to zero-sum thinking: "our" gain is "their" loss. Trust between national security organizations is extremely low. The prevailing assumption is that everyone is a potential enemy and that agreements mean little unless congruent with immediate interests. If the Pentagon had been put in charge of negotiating an ozone protocol, we might still be stockpiling chloroflourocarbons as a bargaining chip. Conventional national security organizations also have short time horizons. The pervasive tendency for national security organizations to discount the future and pursue very near-term objectives is a poor model for achieveing environmental sustainability.

Finally, and perhaps most importantly, privileging national identity and security collides directly with worldviews and identities supportive of sustainable environmental practices. The "nation" is not an empty vessel or blank slate waiting to be filled or scripted, but is instead profoundly linked to "us versus them" thinking. The tendency for people to identify themselves with various tribal and kin groupings is as old as humanity. However, in the last century and a half the sentiment of nationalism, amplified and manipulated by mass media propaganda techniques, has been an integral part of totalitarianism and militarism. Nationalism means a sense of "us versus them" of the insider versus the outsider, of the compatriot versus the alien. The stronger the nationalism, the stronger this cleavage, and the weaker transnational bonds. Nationalism reinforces militarism, fosters prejudice and discrimination, and feeds the quest for "sovereign" autonomy.

In contrast, with environmental problems "we have met the enemy and they are us," as the comic strip figure Pogo aptly observed. As noted earlier, existing "us" versus "them" groupings in world politics match very poorly the causal patterns of environmental problems. At its most basic level, the environmental problem asks us to redefine who "we are" and who "us" encompasses. A central thesis of environmental political thought is that coping with global problems and new forms of ecological interdependence requires replacing or supplementing national with other forms of group identity. Privileging the nation directly conflicts with the "one world" and "global village" sensibility of environmental awareness. Framing environmental problems as threats to "national security" risks undercutting the globalist and common fate understanding of the situation and the sense of world community that are necessary to solve many environmental problems. The resolution of many global environmental problems requires extensive international cooperation, but fueling the fires of nationalist sentiment and identification exacerbates barriers to cooperation. In short, if environmental concerns are wrapped in national flags, the "whole earth" sensibility at the core of environmental awareness will be smothered.

If environmental degradation were to be widely understood as a national security problem, there is also a danger that the citizens of one country would feel much more threatened by the pollution from other countries than by the pollution created by their fellow citizens. This could increase international tensions, make international accords more difficult to achieve, and divert attention from internal clean-up. For example, Americans could become much more concerned about deforestation in Brazil than in halting and reversing North American deforestation. Taken to an absurd extreme—as national security threats sometimes are—seeing environmental degradation in a neighboring country as a national security threat could evoke interventions and armed conflicts.

Americans have been leaders in defining the environmental paradigm and advancing its program, and the particular experiences of the United States heavily shape them. The role of state and nation in the United States has been anomalous, and helps explain the relative blindness of American environmental security analysts to the risks of the environmental security motivational strategy. The American Constitutional order imposes an elaborate system of constraints on the accumulation of centralized state power and the solving of collective action problems. In the twentieth century these restraints have impeded responses to the problems of industrialism domestically and globalization internationally, and external military threats have been catalysts to state-building in the United States. Given this, Americans seeking to build stronger governmental institutions often have found it expedient to frame their social welfare agendas in the terms of national security.[33]

But in much of the rest of the world, too much rather than too little state power threaten public security and liberty. Similarly, nationalism in the United States has centered upon civic rather than ethnic identities, and these liberal identity claims have countered fractious ethnicities and moderated conflicts.[34] Context shapes connotation, and evoking "national security" in the United States has far different and more innocent implications than it does in much of the rest of the world.

Fortunately, environmental awareness need not depend upon co-opted national security thinking. Integrally woven into ecological concerns are a powerful set of interests and values—most notably human health and property values, religions and ethics, and natural beauty and concern for future generations. Efforts to raise awareness of environmental problems can thus connect directly with these strong, basic, and diverse human interests and values as sources of motivation and mobilization. Far from needing to be bolstered by national security mindsets, a "green" sensibility can make strong claim to being the master metaphor for an emerging postindustrial civilization. Instead of attempting to gain leverage by appropriating national security thinking, environmentalists can gain much more political leverage by continuing to develop and disseminate this immensely rich and powerful worldview.

Earth Nationalism

Transposing existing national security thinking and approaches to environmental politics is likely to be both ineffective, and to the extent effective, counterproductive. But the story should not end with this negative conclusion. Fully grasping the ramifications of the emerging environmental problems requires a radical rethinking and reconstitution of many of the major institutions of industrial modernity, including the nation. The nation and the national, as scholars on the topic emphasize, are complex phenomena because so many different components of identity have become conflated with or incorporated into national identities. Most important in Western constructions of national identity have been ethnicity, religion, language, and war memories. However, one dimension of the national—identification with place—has been underappreciated, and this dimension opens important avenues for reconstructing identity in ecologically appropriate ways. Identification with a particular physical place, what geographers of place awareness refer to as "geopiety" and "topophilia," has been an important component of national identity.[35] As Edmund Burke, the great philosopher of nationalism, observed, the sentimental attachment to place is among the most elemental wide-spread and powerful of forces, both in humans and in animals. In the modern era the nation-state has sought to shape and exploit this sentimental attachment.

With the growth of ecological problems, this sense of place and threat to place takes on a new character. In positing the "bioregion" as the appropriate unit for political identity, environmentalists are recovering and redefining topophilia and geopiety in ways that subvert the state-constructed and state-supporting nation. Whether the bioregion is understood as a particular locality defined by ecological parameters, or the entire planet as the only naturally autonomous bioregion, environmentalists are asserting what can appropriately be called "earth nationalism."[36] This construction of the nation has radical implications for existing state and international political communities. This emergent earth nationalism is radical both in the sense of returning to fundamental roots, and in posing a fundamental challenge to the state-sponsored and defined concept of nation now hegemonic in world politics. It also entails a powerful and fresh way to conceptualize environmental protection as the practice of national security.

III. Violent Conflict and Interstate War

Many are attracted to conceptualizing environmental concerns as a national security threat because they anticipate that environmental scarcities and change will stimulate conflict, violence, and interstate war. States often fight over what they value, particularly if related to security. If states begin to be much more concerned with resource and environmental degradation, particularly if they think environmental decay is a threat to their national security, then states may wage wars over resources and pollution. Much of the recent literature on the impacts of climate change upon world politics predicts conflict and violence.[37] As Arthur Westing has asserted: "Global deficiencies and degradation of natural resources, both renewable and nonrenewable, coupled with the uneven distribution of these raw materials, can lead to unlikely—and thus unstable—alliances, to national rivalries, and, of course, to war."[38] In emphasizing such outcomes, environmental security analysts join realist international relations theorists in characterizing international political life as particularly prone to conflict and violence.

To analyze fully the prospects for violent outcomes is a vast and uncertain undertaking.[39] Because there are nearly two hundred independent states, and because resource and environmental problems are diverse and not fully understood, generalizations are hazardous and are likely to have important exceptions. To assess the prospects for resource and pollution wars, I will first make several general points about the methodological weaknesses of recent studies on environmental conflict, consider several overall features of contemporary world politics that make such conflict unlikely, and then examine more closely the six major scenarios for environmental conflict most discussed by environmental security analysts. In general I argue that studies on environmental conflict are deeply flawed in their methodology, that important features of world politics make interstate violence and war much less likely than environmental security analysts suggest, and that these doubts are supported by a balanced consideration of the most frequently discussed scenarios.

Most of the recent works on environmental conflict and violent change suffer from important methodological problems which cast serious doubt on their disturbing conclusions. Many studies on environmental conflict purport to have found trends in the frequency with which environmental scarcities produce conflict. However, it is only possible to find a trend after comparing the historical frequency of conflict against the possible cases of environmental scarcities, making a similar calculation for the present, and then comparing past with present frequency. Unfortunately, most studies on environmental conflict and change do not make historical assessments, and even more alarmingly, fail to consider or even profile the entire set of contemporary cases of environmental scarcity and change to ascertain how many of them have resulted in violent conflict. A second methodological problem is that analysts of environmental conflict do not systematically consider the ways in which environmental scarcity or change can stimulate cooperation. This lacuna is particularly glaring because analysts typically advocate more cooperation as a response to the scarcities and changes they identify or foresee. Many studies also do not control for other sources of conflict.

Another major limitation of most studies on environmental conflict is that they rarely consider the character of the overall international system in assessing the prospects for conflict and violence. Of course, it is impossible to analyze everything at once, but conclusions about conflictual outcomes are premature until the main features of the world political system are factored in. The frequency with which environmental scarcity and conflict will produce violent conflict, particularly interstate wars, is profoundly shaped by six features of contemporary world politics: (1) the prevalance of capitalism and the extent of international trade; (2) the existence of numerous functional international organizations, nongovernmental organizations and epistemic communities; (3) highly developed state-system institutions; and (4) the existence of nuclear weapons; (5) the widespread diffusion of conventional weaponry; and (6) the influence of a hegemonic coalition of liberal constitutional democracies. These deeply rooted material and institutional features of the contemporary world order greatly reduce the likelihood that environmental scarcities and change will lead to interstate violence (see figure 8.1).

FEATURES OF THE CONTEMPORARY WORLD-SYSTEM	IMPLICATIONS FOR ENVIRONMENTAL CONFLICT SCENARIOS
Prevalence of capitalism and extensive international trade	Increased incentives and opportunities for economic development and increased efficiency in resource use; increased costs of conflict Reduced incentives for territorial conquest
Numerous functional international organizations, nongovernmental organizations, and episitemic communities	Diffusion of technical and organizational capabilities; constituency for functional cooperative solutions
Highly developed state-system institutions (diplomacy, alliances, and international law)	Conflict mediation and resolution capacities; regional and global antiaggression norms
Nuclear weapons	Conflict between great powers prohibitively costly; incentives for mutual regulation of military capability
Widespread diffusion of conventional weaponry	Costs of conflict high; conquest of resisting populations difficult
Hegemonic coalition of liberal constitutional democracies	Reduced interstate military rivalry and violent conflict

Figure 8.1 The Contemporary World-System and Environmental Conflict Scenarios

Resource Wars

The hypothesis that states will begin fighting each other as natural resources are depleted and degraded seems intuitively accurate. The popular metaphor of a lifeboat adrift at sea with declining supplies of clean water and rations suggests there will be fewer opportunities for positive-sum gains between actors as resource scarcity grows. Many fears of resource war are derived from the cataclysmic world wars of the first half of the twentieth century. Influenced by geopolitical theories that emphasized the importance of land and resources for great power status, Adolf Hitler fashioned Nazi German war aims to achieve resource autonomy.[40] The aggression of Japan was directly related to resource goals: lacking indigenous fuel and minerals, and faced with a slowly tightening embargo by the Western colonial powers in Asia, the Japanese invaded Southeast Asia for oil, tin, and rubber.[41] Although the United States had a richer resource endowment than the Axis powers, fears of shortages and industrial strangulation played a central role in the strategic thinking of American elites about world strategy.[42] During the Cold War, the presence of natural resources in the Third World helped turn this vast area into an arena for East-West conflict.[43] Given this record, the scenario of conflicts over resources playing a powerful role in shaping international order should be taken seriously.

However, there are three strong reasons for concluding that the familiar scenarios of resource war are of diminishing plausibility for the foreseeable future. First, the robust character of the world trade system means that states no longer experience resource dependency as a major threat to their military security and political autonomy. During the 1930s, the collapse of the world trading system drove states to pursue economic autarky, but the resource needs of contemporary states are routinely met without territorial control of the resource source. As Ronnie Lipschutz has argued, this means that resource constraints are much less likely to generate interstate violence than in the past.[44]

Second, the prospects for resource wars are diminished by the growing difficulty that states face in obtaining resources through territorial conquest. Although the invention of nuclear explosives has made it easy and cheap to annihilate humans and infrastructure in extensive areas, the spread of conventional weaponry and national consciousness has made it very costly for an invader, even one equipped with advanced technology, to subdue a resisting population, as France discovered in Indochina and Algeria, the United States in Vietnam, and the Soviet Union in Afghanistan.[45] At the lower levels of violence capability that matter most for conquering and subduing territory, the great powers have lost effective military superiority and are unlikely soon to regain it.

Third, nonrenewable resources are, contrary to intuitive logic, becoming less economically scarce. There is strong evidence that the world is entering what H. E. Goeller and Alvin M. Weinberg have labeled the "age of substitutability," in which industrial technology is increasingly capable of fashioning ubiquitous and plentiful earth materials such as iron, aluminum, silicon, and hydrocarbons into virtually everything needed by modern societies.[46] The most striking manifestation of this trend is that prices for virtually every raw material have been stagnant or falling for the last two decades despite the continued growth in world economic output. In contrast to the expectations widely held during the 1970s that resource scarcity would drive up commodity prices to the benefit of Third World raw material suppliers, prices have fallen.[47]

Water and Oil

General features of contemporary world politics suggest that resource war scenarios are not very plausible, but two difficult cases—water and oil—warrant more specific attention.

Fresh water is a vital resource that is scarce and becoming scarcer in many parts of the world. In desert and semiarid regions that make up over a third of the earth's land area, water scarcities exert an overwhelming influence because without fresh water these lands are unable to support much life, human or otherwise. In water-deficient regions, violent conflicts over access to supplies of fresh water have occurred since the beginning of recorded history. The potential for conflict is further exacerbated by the fact that many important rivers flow across the lands of many countries, and interstate borders divide underground aquifers.

Given these realities, it is not surprising that "water wars" have been one of the most frequently hypothesized form of resource wars, particularly in the Middle East, a water-scarce region with particularly volatile and violent political relations.[48] Here researchers can point to extensive anecdotal evidence that conflicts over water will cause wars, such as former Egyptian President Anwar Sadat's prediction that Egypt's next war will be over water, or the threats that

Turkey has made to cut off the flow of water in the Euphrates at Iraq's expense. The hydrologist Peter Gleick, one of the most prolific environmental security analysts, claims to have found a "trend toward more conflict" and a "disturbing trend toward the use of force in resource-related disputes."[49] However, this conclusion is not really supported by evidence, and is flawed by the methodological problems mentioned earlier. Asserting a trend is unwarranted because Gleick does not examine the past incidence of water conflict in order to compare the number of actual cases of violent conflict against the possible universe of such cases. Gleick's analysis of the present and future is also anecdotal, and the assertion of a trend is not backed by a comparison of the actual violent water conflicts against the possible ones. Indeed, much of Gleick's analysis hinges on the large number of jointly shared water resources and the importance riparian states have attached to water resources. However, the almost complete absence of violent conflict in these cases suggests that violent conflict over scarce water is a very rare, nearly nonexistent phenomenon in contemporary world politics.

Furthermore, proponents of the "water war" scenario fail to consider the ways in which water scarcity has stimulated cooperation and provided disincentives to violent conflict. Precisely because so many rivers are international, their development requires interstate cooperation.[50] States seeking to gain the benefits of jointly owned water resources are forced to cooperate in their development. There are many important examples of such cooperation that have already occurred, perhaps most notably the Parana River in South America, which Brazil, Paraguay, and Argentina have cooperatively developed. Furthermore, once dams and other extensive infrastructure have been built, interstate violence becomes an increasingly costly option. The plausibility of using the "water weapon" by closing off the flow of water to the detriment of down-stream users is diminished due to the vulnerability of dams to military attack. Some analysts have pointed to the vulnerability of the Aswan High Dam on the Nile to possible Israeli attack as a factor in inducing Egypt to make peace with Israel.[51] Also, because the political relations of the Middle East are so volatile and violent, it is unwise to extrapolate a global trend from largely hypothetical developments in this one region.

Finally, much of the "scarcity" of water projected in many parts of the world presumes the continued existence of highly inefficient use of water and of large subsidies that hide the real economic cost of water usage. As many analysts have pointed out, the introduction of economically rational pricing for water would eliminate much of the projected water "scarcity" with far less social disruption and cost than military aggression to acquire additional supplies.[52]

Oil is the second "hard case" for the critic of resource war scenarios. The stakes in the Persian Gulf War of 1990–91 were of concern to the world community because of Kuwait's extensive oil reserves as well as because of the particularly illegitimate character of the Iraqi aggression and its implications for the security of small states. War over oil remains a real possibility because the Persian Gulf region contains two-thirds of the world's proven oil reserves, and because many of the states in this region are politically unstable and militarily weak relative to their neighbors. But this region is exceptional. With the possible exception of the tiny country of Brunei in Southeast Asia, it is difficult to locate other examples of states that are as oil rich, population poor, and militarily weak in such proximity to militarily powerful states as the Persian Gulf emirates are vis-a-vis Iraq and Iran. Furthermore, the swift and decisive response of nearly the entire international community to Iraqi aggression is likely to deter other states contemplating similar aggressions.

Power Imbalances

Environmental degradation also may affect interstate relations and cause war by altering the relative power capacities of states. Changes in relative power position can contribute to wars, either by tempting a rising state to aggress upon a declining state or by inducing a declining state to attack a rising state before its relative power declines any further. Abundant support for such scenarios can be drawn from history, and international relations scholars have extensively studied such phenomenon.[53]

Alterations in the relative power of states are unlikely to lead to war as readily as the lessons of history suggest because economic power and military power are not as tightly coupled as in the past. The relative economic power position of major states such as Germany and Japan has changed greatly since the end of World War II. But these changes, while requiring many complex adjustments in interstate relations, have not been accompanied by war or the threat of war. In the contemporary world, whole industries rise, fall, and relocate, often causing quite substantial fluctuations in the economic well-being of regions and peoples, without producing wars. There is no reason to believe that changes in relative wealth and power positions caused by the uneven impact of environmental degradation would be different in their effects.

Part of the reason for this loosening of the link between economic and military power has been the nuclear revolution, which has made it relatively cheap for the leading states to deploy staggering levels of violence capacity. Given that the major states field massively oversufficient nuclear forces at the cost of a few percent of their GDP, environmentally induced economic decline would have to be extreme before their ability to field a minimum nuclear deterrent would be jeopardized. A stark example of this new pattern is the fact that the precipitious decline in Russia's economy and defense spending in the 1990s has not diminished Russia's ability to deter great power attack.

Pollution Spillover Wars

A third possible route from environmental degradation to interstate conflict and violence is pollution across interstate borders. It is easy to imagine situations in which one country upstream and upwind another dumps an intolerable amount of pollution on a neighboring country, causing the injured country to attempt to pressure and coerce the source country to eliminate its offending pollution. Fortunately for interstate peace, strongly asymmetrical and significant environmental degradation between neighboring countries is relatively rare. The more typical situation has pollution harming groups in neighboring countries as well as other groups in their own countries. This creates complex sets of winners and losers, and thus a complex array of potential intrastate and interstate coalitions. In general, the more such interactions are occurring, the less likely it is that a persistent, significant, and highly asymmetrical pollution "exchange" will occur. The diverse channels of interdependence in the contemporary world, particularly among the industrialized countries, makes it unlikely that intense cleavages of environmental harm will match interstate borders, and at the same time not be compensated and complicated by other military, economic, or cultural interactions.[54] Resolving such conflicts will be a complex and messy affair, but they are unlikely to lead to war.

Struggles over the Global Commons

There are also conflict potentials related to the global commons. Many countries contribute to environmental degradation of the global commons, and many countries are harmed, but because the impacts are widely distributed, no one country has an incentive to act alone to solve the problem. Solutions require collective action, and with collective action comes the possibility of the "free rider."[55] In the case of a global agreement to reduce carbon dioxide emissions to reduce the threat of global warming, if one significant polluter were to resist joining the agreement, with the expectation that the other states would act to reduce environmnetal harms to a tolerable level, the possibility arises that the states sacrificing to reduce emissions would attempt to coerce the free rider into making a more significant contribution to the effort.

It is difficult to judge this scenario because we lack examples of this phenomenon on a large scale. Free-rider problems may generate severe conflict, but it is doubtful that states would find military instruments useful for coercion and compliance. For example, if China or Russia decided not to join a global warming agreement, it seems unlikely that the other major states would really go to war with such powerful states. Overall, any state sufficiently industrialized to be a major contributor to the carbon dioxide problem is likely to present a very poor target for military coercion.

Impoverishment, Authoritarianism, and War

In the fifth environmental conflict scenario, increased interstate violence results from internal turmoil caused by declining living standards. Most observers agree that the basic source of environmental problems is the modern success in producing wealth, but disagree about the future. Cornucopians argue that new technology, institutional reform, new attitudes, and more efficient capital investment can largely solve the environmental problem without compromising high living standards. But Neo-Malthusians are more pessimistic, asserting that the great wealth generated since the beginning of the industrial revolution cannot be sustained. In this view, the only way to prevent ecological collapse (and the resulting economic collapse) is through radically lower standards of living.

In the Neo-Malthusian scenario, the consequences of economic stagnation for politics and society are likely to be significant and largely undesirable. Although the peoples of the world could live peacefully at lower standards of living, reductions of expectations to conform to these new realities will not come easily. Faced with declining living standards, groups at all levels of affluence are likely to resist reductions in their standard of living by pushing the deprivation onto other groups. Class relations will be increasingly "zero sum games" producing class war and revolutionary upheavals. Faced with these pressures, liberal democratic and free market systems would increasingly be replaced by authoritarian governments capable of maintaining minimum order.[56]

The international consequences of these domestic changes may be increased conflict and war. If authoritarian regimes are more war-prone because of their lack of democratic control, and if revolutionary regimes are war-prone because of their ideological fervor and lack of socialization into international norms and processes, then a world political system containing more such states is likely to be more violent. The historical record from previous economic depressions supports the general proposition that widespread economic stagnation and unmet economic expectations contributes to international conflict.

Although initially compelling, this scenario has flaws as well. First, the pessimistic interpretation of the relationship between environmental sustainability and economic growth is arguably based on unsound economic theory. Wealth formation is not so much a product of cheap natural resource availability as of capital formation via savings and more efficient ways of producing. The fact that so many resource-poor countries, like Japan, are very wealthy, while many countries with more extensive resource endowments are poor, demonstrates that there is no clear and direct relationship between abundant resource availability and economic well-being. Environmental constraints require an end to economic growth based on growing raw material through-puts, rather than an end to growth in the output of goods and services.

Second, even if economic decline does occur, interstate conflict may be dampened, not stoked. In the Neo-Malthusian scenario, domestic political life is an intervening variable connecting environmentally induced economic stagnation with interstate conflict. How societies respond to economic decline may in large measure depend upon the rate at which such declines occur. A compensating factor here is the possibility that as people get poorer they will be less willing to spend increasingly scarce resources for military capabilities. In this regard, the experience of economic depressions over the last two centuries may not be relevant, because such depressions were characterized by underutilized production capacity and falling resource prices. In the 1930s increased military spending had a stimulative effect, but in a world in which economic growth had been retarded by environmental constraints military spending would exacerbate the economic problem.

State Collapse and Internal Conflict

The sixth, and most plausible, scenario for environmental conflict centers upon internal political conflict arising from environmental scarcities and change. This scenario has been emphasized by Thomas Homer-Dixon and the other members of his research team.[57] Although this research is sophisticated relative to much of the environmental security literature, it also suffers from the methodological problems discussed earlier, both absence of historical comparison and a failure to examine cases of environmental change that either did not lead to violent conflict or that stimulated cooperative arrangements. As such it neither demonstrates trends nor verifies propositions on causal relationships. Nevertheless, the work of this research project marks a considerable advance and makes three important contributions. First, it advances the field's conceptualization of the processes and pathways through which environmental scarcities, particularly in degraded or depleted renewable resources of forests, fisheries, and soils, can cause or exacerbate social conflicts to the point of violence. Although such process tracing falls short of the verification of causal propositions, it sheds important light on the ways in which social and natural systems can interact. Second, its careful case studies of ecological and social developments has moved the field beyond the reliance upon anecdotal evidence, and thus helped lay the ground work for further analysis. Third, it should be noted that Homer-Dixon and his colleagues have not found much merit in the scenarios for interstate conflict, and thus have largely corroborated the doubts raised by critics of the first wave of environmental security work.

Assuming that environmental degradation can lead to internal turmoil and state collapse, what are the international ramifications? Should some areas of the world suffer this fate, the impact of this outcome on international order may not be very great. If a particular country, even a large one like India or Brazil, were to disintegrate, among the first casualties would be the capacity of the industrial and governmental structure to wage and sustain interstate conventional war. As Bernard Brodie observed about the modern era, "the predisposing factors to military aggression are full bellies, not empty ones."[58] The poor and wretched of the earth may be able to deny an outside aggressor an easy conquest, but they are themselves a minimal threat to outside states. Offensive war today requires complex organizational skills, specialized industrial products, and surplus wealth.

In the contemporary world connectivity is high, but not tightly coupled. Regional disasters of great severity may occur, with scarcely a ripple in the rest of the world. Idi Amin drew Uganda back into savage darkness, the Khmer Rouge murdered an estimated two million Cambodians, and the Sahara has advanced across the Sahel without the economies and political systems of the rest of the world being much perturbed. Indeed, many of the world's citizens did not even notice.

In summary, the case for thinking environmental degradation will cause interstate violence is much weaker than commonly thought. In part this is because of features of the international system unrelated to environmental issues. Although many conflict scenarios draw analogies from historical experience, they fail to take into account the important ways in which the contemporary interstate system differs from earlier ones. Military capability sufficient to make aggression prohibitively costly has become widely distributed, making even large shifts in the relative power potential of states less likely to cause war. Interstate violence seems to be poorly matched as a means to resolve many of the conflicts that might arise from environmental degradation. The vitality of the international trading system and the more general phenomenon of complex interdependence also militate against violent interstate outcomes. The result is a world system with considerable resiliency and enough "rattle room" to weather significant environmental disruption without large-scale violent interstate conflict.

Conclusions

The degradation of the natural environment upon which human well-being depends is a problem with far-reaching significance for all human societies. But this problem has little to do with the national-security-from-violence problem that continues to afflict politics. Not only is there little in common between the causes and solutions to these two problems, but the nationalist and militarist mindsets closely associated with "national security" thinking directly conflict with the core of the emerging environmentalist worldview. Harnessing these sentiments for a "war on pollution" is a dangerous and probably self-defeating enterprise. And fortunately, the prospects for resource and pollution wars are not as great as often conjured by environmentalists.

Overall, the pervasive recourse to "national security" paradigms to conceptualize the environmental problem represents a profound and disturbing failure of political imagination. If the nation-state enjoys a more prominent status in world politics than its competence and accomplishments warrant, then it makes little sense to emphasize the links between it and the emerging problem of the global habitability.[59] Nationalist sentiment and the war system have a long-established logic and staying power that are likely to defy any rhetorically conjured redirection toward benign ends. The movement to preserve the habitability of the planet for future generations must directly challenge the power of state-centric nationalism and the chronic militarization of public discourse. Environmental degradation is not a threat to national security. Rather, environmentalism is a threat to the conceptual hegemony of state-centered national security discourses and institutions. For environmentalists to dress their programs in the blood-soaked garments of the war system betrays their core values and creates confusion about the real tasks at hand.

-------------------------------- NOTES --------------------------------

1. For the extent of this dependence, see Gretchen C. Daily, ed., *Nature's Services: Societal Dependence on Natural Ecosystems* (Washington, D.C.: Island Press, 1997).

2. Lester Brown, "Redefining National Security," Worldwatch Paper number 14, October 1977.

3. Richard H. Ullman, "Redefining Security," *International Security* vol. 8, no. 1 (Summer 1983), pp. 129–53.

4. Arthur H. Westing, "Global Resources and International Conflict: An Overview," in Westing, ed., *Global Resources and Environmental Conflict: Environmental Factors in Strategic Policy and Action* (New York: Oxford University Press, 1986).

5. Patricia M. Mische, "Ecological Security and the Need to Reconceptualize Sovereignty," *Alternatives* vol. 14 (1989), pp. 389–427.

6. Michael Renner, "National Security: The Economic and Environmental Dimensions," Worldwatch Paper Number 89, May 1989; Jessica Tuchman Mathews, "Redefining Security," *Foreign Affairs* vol. 68, no. 2 (Spring 1989), pp. 169–71; Norman Myers, "Environmental Security," *Foreign Policy* no. 74

(1989), pp. 23–41; and Gwyn Prins, ed., *Threats without Enemies: Facing Environmental Insecurity* (London: Earthscan, 1993).

7. Robert Kaplan, "The Coming Anarchy," *The Atlantic Monthly* vol. 273, no. 2 (February 1994), pp. 44–76; and Robert Kaplan, *The Ends of the Earth: A Journey to the Frontiers of Anarchy* (New York: Vintage Books, 1996).

8. For accounts of the influence of "environmental security" in the Clinton administration, see Jeremy D. Rosner, "Is Chaos America's Real Enemy? The Foreign Policy Splitting Clinton's Team," *Washington Post Outlook*, September 14, 1994, p.1; and Jeremy D. Rosner, "The Sources of Chaos," *The New Democrat*, November 20–22, 1994.

9. For a description and defense, see the chapter by Ken Butts in this volume.

10. Daniel Deudney, "The Case against Linking Environmental Degradation and National Security," *Millennium* vol. 19 (1990), pp. 461–76; and Daniel Deudney, "Environmental Security: Muddled Thinking," *Bulletin of the Atomic Scientists* vol. 47, no. 3 (1991), pp. 22–28.

11. Mathias Finger, "The Military, the Nation State, and the Environment" *The Ecologist*, vol. 21, no. 5 (1991); Simon Dalby, "Ecopolitical Discourse: 'Environmental Security' and Political Geography," *Progress in Human Geography* vol. 16 (1992), pp. 503–22; Ken Conca, "In the Name of Sustainability: Peace Studies and Environmental Discourse," in Jyrki Kakonen, ed., *Green Security or Militarized Environment* (Aldershot: Dartmouth, 1994); and Marc A. Levy, "Is the Environment a National Security Issue?" *International Security* vol. 20, no. 2 (Winter 1995), pp. 35–62.

12. For a succinct overview, see: Geoffrey D. Dabelko and P. J. Simmons, "Environment and Security: Core Ideas and U.S. Government Initiatives," *SAIS Review* vol. 17, no. 1 (1997), pp. 127–46.

13. For a discussion of the interplay between conceptual, terminological, and political change, see Quentin Skinner, "Language and Political Change," and James Farr, "Understanding Political Change Conceptually," in Terence Ball et al., eds., *Political Innovation and Conceptual Change* (Cambridge, U.K.: Cambridge University Press, 1989), pp. 6–23, 24–49.

14. For a particularly lucid and well-rounded discussion of security, the state, and violence, see Barry Buzan, *People, States, and Fear: The National Security Problem in International Relations* (Chapel Hill: University of North Carolina Press, 1983).

15. Clive Ponting, *A Green History of the World: The Environment and the Collapse of Great Civilizations* (New York: St. Martin's, 1992).

16. SIPRI, *Warfare in a Fragile World: Military Impact on the Human Environment* (London: Taylor and Francis, 1980)

17. T. W. Hawley, *Against the Fires of Hell: The Environmental Disaster of the Gulf War* (New York: Harcourt Brace Jovanovich, 1992).

18. Paul R. Ehrlich, Carl Sagan, Donald Kennedy, and Walter Orr Roberts, *The Cold and the Dark: The World After Nuclear War* (New York: Norton, 1984).

19. For the best overall analysis, see Michael Renner, "Assessing the Military's War on the Environment," in Lester Brown, ed., *State of the World 1991*, (New York: Norton, 1991) pp. 132–52.

20. Zhores A. Medvedev, *The Nuclear Disaster in the Urals* (New York: Norton, 1990).

21. William James, "The Moral Equivalent of War," *Memories and Studies* (New York and London: Longmans, Green & Co., 1911).

22. Daniel Deudney, "The Case against Linking Environmental Degradation and National Security," *Millennium* vol. 19 (1990), pp. 461–76; and Ken Conca, "In the Name of Sustainability: Peace Studies and Environmental Discourse," in Jyrki Kakonen, ed., *Green Security or Militarized Environment* (Aldershot, U.K.: Dartmouth, 1994).

23. Jyrki Kakonen, ed., *Green Security or Militarized Environment* (Aldershot, U.K.: Dartmouth, 1994).

24. Michael Howard, "War and the Nation-State," *Daedalus* vol. 108, no. 4 (Fall 1979), pp. 101–10.

25. Charles Tilly, "War Making and State Making as Organized Crime," in Peter Evans, Dietrich Rueschemeyer and Theda Skocpol, eds., *Bringing the State Back In* (Cambridge, U.K.: Cambridge University Press, 1985), pp. 169–91.

26. Nancy Peluso, *Rich Forests, Poor People* (Berkeley: University of California Press, 1993); and Charles S. Wood and Marianne Schmink, "The Military and the Environment in the Brazilian Amazon," *Journal of Political and Military Sociology* vol. 21 (Summer 1993), pp. 81–105.

27. Jason Clay, "Resource Wars: Nation and State Conflicts of the Twentieth Century,"in Barbara Rose Johnston ed, *Who Pays the Price? The Sociocultural Context of Environmental Crisis* (Washington, D.C.: Island Press, 1994).

28. James Der Derian, "Security," in Ronnie Lipschutz, ed., *On Security* (New York: Columbia University Press, 1995).

29. Rudolph J. Rummel, "Democracy, Power, Genocide and Mass Murder," *Journal of Conflict Resolution* vol. 39, no. 3 (1995), pp. 3–26.

30. Eric K. Stern, "Bringing the Environment In: The Case for Comprehensive Security," *Cooperation and Conflict* vol. 30, no. 3 (1995), pp. 211–37; and Norman Myers, *Ultimate Security: The Environmental Basis of Political Stability* (New York: Norton, 1993).

31. Michael Ross Fowler and Julie Marie Bunck, *Law, Power, and the Sovereign State: The Evolution and Application of the Concept of Sovereignty* (University Park: Pennsylvania State University Press, 1995).

32. Joseph A. Yager, "The Energy Battles of 1979," in Craufurd D. Goodwin, ed., *Energy Policy in Perspective* (Washington, D.C.: Brookings, 1982), pp. 601–36.

33. William E. Leuchtenburg, "The New Deal and the Analog of War," in John Braeman, ed., *Change and Continuity in Twentieth-Century America* (Columbus: Ohio University Press, 1964).

34. For the classic statement of this view, see Hans Kohn, *American Nationalism: An Interpretative Essay* (New York: Macmillan, 1957).

35. Yi-Fu Tuan, "Geopiety: A Theme of Man's Attachment to Nature and Place," in David Lowenthal and Nartyn J. Bowden eds., *Geographies of the Mind* (New York: Oxford University Press, 1976); and Yi-Fu Tuan, *Topophilia: A Study of Environmental Perception, Attitudes, and Values* (Bloomington: University of Indiana Press, 1994).

36. Daniel Deudney, "Ground Identity: Nature, Place, and Space in the National," in Yosef Lapid and Friedrich Kratochwil, eds., *The Return of Culture and Identity to International Relations Theory* (Boulder, Colo.: Lynne Reinner, 1995), pp. 129–45.

37. Peter Gleick, "The Implications of Global Climatic Changes for International Security," *Climatic Change* vol. 15 (1989), pp. 309–25; Peter Gleick, "Global Climatic Changes and Geopolitics: Pressures on Developed and Developing Countries," in A. Berger et al., eds., *Climate and Geo-Sciences* (Amsterdam: Kluwar Academic Publisher, 1989); and Peter H. Gleick, "Environment and Security: The Clear Connections," *Bulletin of the Atomic Scientists* vol. 47, no. 3 (1991).

38. Arthur Westing, ed., *Global Resources and International Conflict: Environmental Factors in Strategic Policy and Action*, SIPRI (Oxford: Oxford University Press, 1986).

39. For an extended analysis, see Thomas F. Homer-Dixon, *Environment, Scarcity, and Violence* (Princeton, N.J.: Princeton University Press, 1998).

40. For discussions of resource autarky during the 1930s see Brooks Emeny, *The Strategy of Raw Materials* (New York: Macmillan, 1934); Norman Rich, *Hitler's War Aims: Ideology, The Nazi State, and the Course of Expansion* (New York: Norton, 1973); and William Carr, *Arms, Autarchy, and Aggression: A Study in German Foreign Policy, 1933–1939* (London: Edward Arnold, 1972).

41. James Crowley, *Japan's Quest for Autonomy: National Security and Foreign Policy, 1930–1938* (Princeton, N.J.: Princeton University Press, 1966).

42. Nicholas John Spykman, *America's Strategy in World Politics: The United States and the Balance of Power* (New York: Harcourt, Brace, 1942).

43. Alfred E. Eckes, Jr., *The United States and the Global Struggle for Minerals* (Austin, Tex.: University of Texas Press, 1979).

44. Ronnie D. Lipschutz, *When Nations Clash: Raw Materials, Ideology and Foreign Policy* (Cambridge, Mass.: Ballinger, 1989).

45. Among the most recent versions of the argument that war is of declining viability are Evan Luard, *The Blunted Sword: The Erosion of Military Power in Modern World Politics* (New York: New Amsterdam Books, 1989); and John Mueller, *Retreat from Doomsday: The Obsolescence of Major War* (New York: Basic Books, 1989).

46. H. E. Goeller and Alvin Weinberg, "The Age of Substitutability," *Science*, 1967, vol. 201; and Eric D. Larson, Marc H. Ross, Robert H. Williams, "Beyond the Era of Materials," *Scientific American* vol. 254 (1986), pp. 34–41; and Iddo K. Wernick, Robert Herman, Shekler Govind, and Jesse H. Ausubel, "Materialization and Dematerialization," *Deadalus* vol. 125, no. 3 (Summer 1996), pp. 171–98.

47. H. E. Goeller, "Trends in Nonrenewable Resources," ch. 30 in Julian Simon, ed., *The State of Humanity* (Oxford: Blackwell, 1995), pp. 313–22; and Stephen Moore, "The Coming Age of Abundance," in Ronald Bailey, ed., *The True State of the Planet* (New York: Free Press, 1995), pp. 110–39.

48. Joyce R. Starr and Daniel C. Stoll, eds., *The Politics of Scarcity: Water in the Middle East* (Boulder, Colo.: Westview, 1988); John Bulloch and Adel Darwish, *Water Wars: Coming Conflicts in the Middle East* (London: Victor Collancx, 1993); and Daniel Hillel, *Rivers of Eden: The Struggle for Water and the Quest for Peace in the Middle East* (Oxford: Oxford University Press, 1994).

49. Peter Gleick, "Water and Conflict," occasional paper series of the Project on Environmental Change and Acute Conflict, number 1, 1992.

50. Leif Ohlsson, ed., *Hydropolitics: Conflicts over Water as a Development Constraint* (London: Zed, 1995); Virginia Wheeler, "Cooperation for

Development in the Lower Mekong Basin," *American Journal of International Law* vol. 64 (1970), pp. 594–609; and Dale Whittington and Elizabeth McClelland, "Opportunities for Regional and International Cooperation in the Nile Basin," *Water International* vol. 17 (1992), pp. 144–54.

51. John Waterbury, *Hydropolitics of the Nile Valley* (Syracuse, N.Y.: Syracuse University Press, 1979).

52. Peter Beaumont, "The Myth of Water Wars and the Future of Irrigated Agriculture in the Middle East," *International Journal of Water Resources Development* vol. 10 (1994), pp. 9–22.

53. Robert Gilpin, *War and Change in World Politics* (Cambridge, U.K.: Cambridge University Press, 1981).

54. For analysis of "complex interdependence," see Robert Keohane and Joseph Nye, *Power and Interdependence* (Boston: Little, Brown, 1977).

55. For the classic statement on the "free rider," see Mancur Olson, Jr., *The Logic of Collective Action* (Cambridge, Mass.: Harvard University Press, 1965).

56. For discussion of authoritarian and conflictual consequences of environmentally constrained economies, see Robert Heilbroner, *An Inquiry into the Human Prospect* (New York: Norton, 1974); William Ophuls, *Ecology and the Politics of Scarcity* (San Francisco: Freeman, 1976); Susan M. Leeson, "Philosophical Implications of the Ecological Crisis: The Authoritarian Challenge to Liberalism," *Polity* vol. 11 (1979); and Ted Gurr, "On the Political Consequences of Scarcity and Economic Decline," *International Studies Quarterly* vol. 29 (1985), pp. 51–75.

57. Thomas Homer-Dixon, "Across the Threshold: Environmental Scarcities and Violent Conflict: Evidence from Cases," *International Security* vol. 19, no. 1 (1994).

58. Bernard Brodie, "The Impact of Technological Change on the International System," in David Sullivan and Martin Sattler, eds., *Change and the Future of the International System* (New York: Columbia University Press, 1972), p. 14.

59. For a classic statement of the argument that the nation-state system is overdeveloped relative to its actual problem-solving capacities, see George Modelski, *Principles of World Politics* (New York: Free Press, 1972).

PART III

Case Studies

9

Transboundary Resource
Disputes and Their
Resolution

Miriam R. Lowi

At the first round of Middle East peace talks, held in Moscow in January 1992, the participants formed working groups on five substantive issues of mutual concern—refugees, the environment, economic development, arms control, and water resources. Established within the framework of the multilateral track of the peace process, these working groups were to meet separately from, but when possible, simultaneously with, the bilateral negotiations. The intention was that the multilateral meetings would support and complement the bilateral meetings, and that progress at the technical level in the working groups would inspire the political negotiations, and vice versa. The convenors of the peace conference also hoped that technical agreements could be arrived at and implemented even before a positive conclusion of the bilateral negotiations.

By January 1996, the water resources working group had met seven times: in Vienna in May 1992, Washington in September 1992, Geneva in April 1993, Beijing in October 1993, Muscat in April 1994, Athens in November 1994, and Amman in June 1995. Until the Arafat-Rabin signing of the *Declaration of Principles on Interim Self-Government Arrangements* ("Oslo I") at the White House on 13 September 1993, discussions in the water resources working group had not gone far beyond agreement on the obvious: that there was not enough water in the region, that consumption demand was growing, that water quality was deteriorating rapidly. Lengthy discussions took place on the problem of data collection, and the parties agreed that under more favorable political conditions, it would be important to share data. There it stood.[1]

In the five working groups, significant headway was impeded by the "high politics" conflict. Both sides, but especially the Arabs, were reluctant to approve any cooperative activity in the absence of progress in the bilateral meetings toward a political settlement of the core issue of dispute: the future status of the territories occupied by Israel in June 1967. This is not surprising, given the history of the dispute over the Jordan waters and the various efforts to resolve it.

From the time of the signing of the *Declaration of Principles* until the defeat of the Labour Party government in June 1996, the atmosphere in the water resources working group improved considerably and substantive progress was made. Team members have begun to think in the long term and plan for not only continued interactions, but joint water projects as well. Nonetheless, discussions about regional and basinwide resource management were put off largely because, in the absence of firm political commitment on the part of all parties to the conflict, team members were reluctant to consider the matter seriously. It may be that basinwide accord awaits final status negotiations that culminate in the creation of a Palestinian state and peace treaties between each Arab state and Israel.

This paper considers the linkage between "low politics" and "high politics" in conflict resolution, as it relates to resource scarcity, resource dependence, and the Israeli-Palestinian relationship. Specifically, I focus on Israel's dependence on the subterranean water supply of the West Bank and its implications for a political settlement in the region. I begin by sketching the availability and use of water resources in Palestine/Israel. I demonstrate the degree to which Israel is dependent upon water that originates in occupied territory. A brief overview is provided of the role that both water resources and the West Bank have played in Israeli security thinking. Following that, I outline Israel's West Bank water policies and some of the effects they have had on the local environment. Next, in an effort to relate the water issue to the current Middle East peace process, I discuss the various lessons that can be learned from the history of the Jordan waters dispute about conflict resolution in the Israeli-Arab-Palestinian arena. I conclude with some remarks on the status of the water component of the peace process and prospects for the future.

I. The Jordan River Basin

The Jordan Basin is an elongated valley in the central Middle East. Draining some 18,300 square kilometers, it extends from Mount Hermon in the north to the Dead Sea in the south, and lies within the pre-June 1967 boundaries of Israel, Jordan, Lebanon, and Syria. Its waters drain the land both east and west of the Jordan Valley. Precipitation in the basin ranges from over 1000 mm/yr (millimeters per year) in the north to less than 50 mm/yr in the south, but averages less than 200 mm/yr on both sides of the Jordan River. Much of the basin, therefore, is arid or semiarid, and requires irrigation water for agricultural development.

While the basin extends into four states, about 80 percent of it is located in present-day Israel, Jordan, and the West Bank. It is these lands that are the most dependent upon its waters. Moreover, the political conflict in the region has been intertwined, since its inception, with a dispute over access to the water resources of the Jordan basin.[2]

The West Bank, so named because of its location relative to the Jordan River, covers an area of 5946 square kilometers (including East Jerusalem).[3] It borders the Negev Desert to the south, the Coastal Plain to the west, the Galilee to the north, and the Jordan River and the Kingdom of Jordan to the east. The territory comprises hilly regions in the north, west, and center, valley lands in the east, and desert in the southeast.

Table 9.1 Jordan basin: Principal surface waters (MCM/YR)[a]

RIVER	SOURCE	DIRECTION	POLITICAL CONTROL	DISCHARGE
Hasbani	Lebanon	south to Upper Jordan River	Lebanon pre-67 Israel post-67	138
Banias	Syria	south to Upper Jordan River	Syria pre-67 Israel post-67	121
Dan	Israel	south to Upper Jordan River	Israel	245
Upper Jordan		south to Lake Tiberias	Israel	650
Lower Jordan		south to Dead Sea	Israel/Jordan	1200[b]
Yarmouk	Syria	southwest to Lower Jordan	Syria/Jordan/ Israel	450

a. These figures represent average flow under "normal" climatic conditions and prior to extractions for irrigation and development. Discharge figures for the Upper and Lower Jordan include the contributions of tributaries and other sources. In this chapter, million cubic meters (mcm) is used as the standard measure of water volume. One cubic meter of water equals 1,000 kilograms in weight, or one metric ton.

b. This figure represents the flow into the Dead Sea until the early 1950s, prior to the inception of development schemes in the basin. Today, the flow is no more than 100 mcm per annum. Moreover, the waters of the Lower Jordan are highly saline and cannot be used for most agricultural purposes.

Source: Compiled by author from Charles T. Main, Inc., *The Unified Development of the Water Resources of the Jordan Valley Region* (Boston, Charles T. Main, 1953)

II. Water Resources and Their Utilization

Israel's overall water inventory is limited, unevenly distributed, and subject to rather high, climatically determined fluctuations. The country has a Mediterranean climate, varying from semiarid in the north to arid in the south. More than one-half the area of the country receives an annual average of less than 200 millimeters of rainfall. Water resources are unfavorably located in relation to main areas of demand. While water is most plentiful in the north and northeast, the densest concentrations of population, industry, and irrigable land are in the center of the country and in the coastal plain. There are also temporal problems with the distribution of water. Streamflow and storm-water runoff are at their peak during the winter months, whereas consumption peaks at the height of the irrigation season in July and August. Moreover, drought years and even successions of drought years are not at all uncommon. Finally, the extent of stream flow is small: the annual volume of the Upper Jordan River at the northern tip of Lake Tiberias, Israel's largest exploitable body of surface water, represents less than one-third of the country's total demand. In contrast, approximately two-fifths of the country's total renewable water potential—that is, 850 mcm out of about 1900 mcm—originate from groundwater, located in three principal aquiferous formations.[4] The coastal plain aquifer and the more abundant Yarqon-Taninim aquifer (also called the "mountain" or "western" aquifer) provide the bulk of the groundwater supply [table 9.2].

Because of the rapid increase in population and the marked expansion of irrigated area since 1949, demands on the limited resources of the country have been great. By 1990, Israel's population had quadrupled in size, while the irrigated area increased sevenfold. Total water consumption has increased more than eight times.[5] Since the mid 1970s, in fact, demand for fresh water has exceeded the

Table 9.2 Israel: Fresh water resources (1990)

SOURCE	QUANTITY (MCM/YR)
1. Surface water	
Upper Jordan	580
Yarmouk River and flood water	80–180[a]
2. Groundwater	850
3. Recycled wastewater	360
Total	1870–1970

a. Note that these numbers vary considerably with climatic conditions.

Source: Compiled by author from *Associates for Middle East Research* Data Base; Schwarz 1992; Sofer 1992; Lonergan and Brooks 1994.

country's sustainable annual water yield—the fixed quantity of water available on a yearly basis—while the prospects for the development of yet unexploited conventional fresh water resources remain extremely limited.

During the basinwide drought conditions of the 1980s, supply constraints intensified. Consumption of water in agriculture had to be reduced for the first time in 1986.[6] Since then, the government has been forced to cut back the sector's allocation in order to satisfy urban and industrial demand. Indeed by 1996, overall water consumption had been reduced by about 200 mcm per annum, from 1950 mcm to roughly 1750 mcm. Given supply constraints and population growth projections, trends toward increasing scarcity and the belt-

Table 9.3 Population, water supply, and demand projections

	ISRAEL	WEST BANK	JORDAN
Population			
1990	4.6 mill.[a]	1.8 mill.[b]	3.2 mill.
2020	6.7	4.7	9.8
Water Supply			
1987–91	1950 mcm[c]	650 mcm	900 mcm
drought[d]	1600 [e]	450–550	700–750
2010	2060 [e]		
Water Demand			
1987–91	2100 mcm[a]	125 mcm[f]	800 mcm
2020	2800	530 [f]	1800

a. Includes Jewish settlements in Occupied Territories and Golan Heights.

b. Arab population of West Bank and Gaza Strip.

c. This figure includes some 520 mcm of groundwater originating in the West Bank.

d. Average annual supply of water during drought conditions of 1980s.

e. This figure includes some portion of groundwater originating in the West Bank.

f. Demand for water by the Arab population of the West Bank alone.

Source: Associates for Middle East Research Data Base. [Note that this table was compiled prior to the September 1993 *Declaration of Principles*; hence the figures do not reflect the changes anticipated by that document.]

tightening measures that accompany them are bound to continue.

In the absence of additional sources of surface water, groundwater tends to be overexploited, thereby diminishing its quality and threatening its future availability. Since the sustainable annual yield is a fixed quantity, excess withdrawals by overpumping or depletion of underground reserves constitute an overdraft that could cause irreversible damage. Overpumping lowers the water table, increasing the danger of salt water infiltration. When the reserve of underground flow sinks below a certain level in the coastal aquifer, the interface, or dividing line, between fresh and sea water is drawn upward and causes salination. Israel has been overpumping groundwater sources since 1970.[7] The deteriorating quality of the water supply has become a cause of growing concern.[8]

The critical point about the groundwater sources upon which Israel is dependent is their location and direction of flow. Of the three main aquifer groups, only one is located in Israel proper, beneath the coastal plain. This is the second most abundant of the three. The remaining two originate in occupied territory. The most abundant— the Yarqon-Taninim aquifer—extends along the western foothills of the West Bank. Its natural recharge flows across the 'Green Line' (the 1949 United Nations Armistice Demarcation line) into Israel's coastal plain. The least abundant—those aquifer groups in the northern part of the West Bank—drain an area across the Green Line as well, and discharge into the Bet She'an and Jezre'el Valleys. Both the western and northern basins can be tapped from either side of the Green Line. However, only 5 percent of the combined recharge area of the two water tables is located in Israel proper.[9] Another group of aquifers, the smallest of the West Bank water tables, forms the eastern drainage basin. Its water does not traverse the Green Line; rather, it discharges into the Jordan Valley and hence, cannot be exploited outside the territory.[10] (See figure 9.1.) Thus, between 475 and 490 mcm, or roughly 57 percent of Israel's sustainable annual supply of groundwater and one-quarter of its total renewable fresh water supply, originate in occupied territory.

The water resources of the West Bank consist of surface runoff and groundwater. Of a total availability of about 650 mcm per annum, the consumption of water by the Arab population of the West Bank in the 1980s did not exceed 125 mcm, or 19 percent.[11] Hence, the remainder of about 525 mcm, minus that which was lost to evaporation or surface runoff, represented the quantity that could be exploited by the non-Arab population and/or beyond the West Bank.

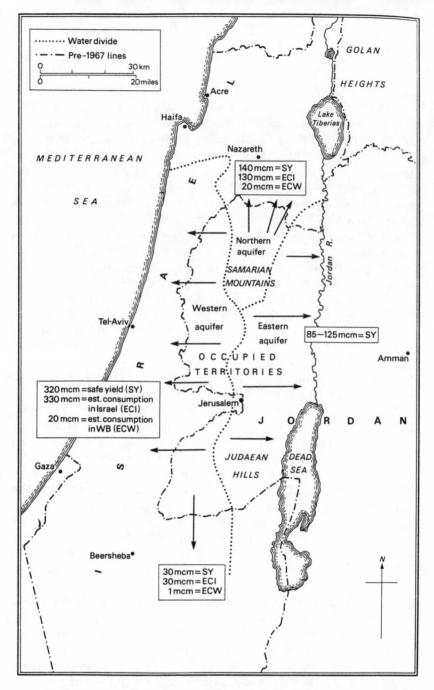

Figure 9.1 In February 1993, an Israeli geographer told this author that Israel was taking about 390 mcm per annum from the western basin; in other words, the aquifer was overexploited by at least 55 mcm per annum.

III. Water and the West Bank in Israeli Security Thinking

The establishment of the state of Israel was, to a considerable degree, the product of the Zionist movement's concerns for the security and survival of world Jewry. European Zionists of the late nineteenth century argued that Jews would continue to be persecuted until they constituted a majority in a territory over which they held sovereign control. Hence, the answer to what was referred to as the "Jewish Question" was the founding of a Jewish national home and the "ingathering of the exiles." This position was given irreversible momentum with the rise of Nazism in central Europe in the 1930s and the extermination of six million Jews.

From the outset of the Zionist movement's endeavors, unrestricted access to water resources was perceived as a nonnegotiable prerequisite for the survival of a Jewish national home. In fact, concerns for the economic viability of a Jewish state in arid Palestine prompted the World Zionist Organization to insist at the Paris Peace Conference on 3 February 1919, that it was "of vital importance not only to secure all water resources already feeding the country, but also to be able to conserve and control them at their sources."[12] The Organization submitted a frontier claim to the conference that included the whole of the Jordan River basin and headwaters, as well as most of the Litani River.[13]

The Zionist movement also considered water to be important insofar as it was part of the "ideology of agriculture" in Zionist thought.[14] By working the land, Jews would be returning to *Eretz Yisrael* (the Land of Israel) in the most literal sense. Moreover, to Socialist Zionism which dominated the movement, the "ideal man" is he who tills the land. And in keeping with socialist doctrine, he who tills the land has rights to it.

In time, however, agriculture—and by extension, water—became related to defense and defense imperatives. Not only was it essential to be on the land in order to lay claim to it, but also, to quote one Israeli interviewed: "Jews must work the land—all the land—because if not, it will be occupied by Arabs. This would be the end of the Jewish state."[15] Agricultural development has remained a national goal, embodying a socially accepted value and dictated by ideology.

Water, because it is an essential ingredient of agriculture, has always been linked in some fashion to Zionists' security-related concerns. Moshe Sharett, a former prime minister of Israel, described the primacy of water resources in the continued survival of the Jewish state:

> Water for Israel is not a luxury; it is not just a desirable and helpful addition to our system of natural resources. *Water is life itself*. It is bread for the nation—and not only bread. Without large irrigation works we will not reach high production levels . . . to achieve economic independence. And without irrigation we will not create an agriculture worthy of the name . . . and without agriculture . . . we will not be a nation rooted in its land, sure of its survival, stable in its character, controlling all opportunities of production with material and spiritual resource.[16]

From the June 1967 war until roughly the Persian Gulf war (1991), the continued occupation of the West Bank remained a basic ingredient of official Israel government rhetoric with regard to national security. Retaining control over the territory was perceived as the solution to the problem of Israel's geographic vulnerability. Within the pre-1967 boundaries, a majority of the population and of big industry inhabit the narrowest portion of the country, the Coastal Plain. This means that there is no strategic depth between the Mediterranean Sea and the Green Line. As a result, Israel is strategically the most vulnerable precisely where it is economically and demographically the strongest.[17] Hence, the dominant view was that Israeli control of the West Bank enhanced the state's capabilities: it offered strategic depth and provided a natural frontier, the Jordan River, which could be more easily defended than could the Green Line. Moreover, by providing more land for the "ingathering of exiles," it served the ideological interests of the state and afforded the possibility of narrowing the margin of quantitative inferiority. Continued occupation of the West Bank also guaranteed the state control over vital water supplies that originate in the territory but are consumed, for the most part, in Israel.

IV. Israel's West Bank Water Policies

Israel's water resources are administered by the Israeli Water Commission, which is under the authority of the Minister of Agriculture. Two other institutions are part of the Water Commission Administration: Mekorot, Israel's national water authority, and Tahal, the Water Planning for Israel Company. Mekorot is responsible for the construction of irrigation and water supply projects, and Tahal, for the overall planning and design of water development schemes. Following the June 1967 war, the water resources of the Occupied Territories were gradually integrated into Israel's water system.[18] They were administered by the Water Commission Administration until September 1995 and the creation, within the framework of the Middle East peace process, of the Israeli-Palestinian Joint Water Committee. (See below, section VII.)

The policy set forth by the commissioner allowed West Bank Arabs a total consumption of 125–130 mcm per annum. Stringent measures were adopted to ensure that this policy was respected. First, no wells could be drilled on the West Bank without permission from the Civil Administration, Tahal, and Mekorot. No Palestinian Arab individual or village received permission to drill a new well for agricultural purposes, nor to repair one that was close to an Israeli well.[19] Sinking wells on the mountain ridge, the location of the Yarqon-Taninim aquifer, was strictly forbidden to anyone but Mekorot.[20] Occasionally, permission was granted for the drilling of wells designed for household use.

Second, for the agricultural activities of West Bank Arabs, only "existing uses" of water were recognized: uses which had existed in 1967–68.[21] Thus, water allocations to Arab agriculture remained at their 1968 level of 100 mcm, with a slight margin for growth. Third, the technology for deep drilling and rock drilling—which would be required in the western basin, at least—remained in Israeli hands. Fourth, West Bank Arabs were not allowed to use water for farming after four o'clock in the afternoon.[22] Fifth, strict limits were placed on the amount of water that could be pumped annually from each well; meters fixed to wells monitored the amounts extracted. In 1983, the total quantity of well water permitted to the Arab population for agricultural purposes did not exceed the 1967–68 level of 38 mcm.[23] Finally, whereas the water utilized by Jewish settlers was heavily subsidized by the state, West Bank Palestinians received no subsidy at all;[24] they paid a higher price per cubic meter of water than did Jewish settlers. It has been estimated that in 1990, Palestinians were paying as much as six times more for water than the settlers.[25]

Israeli measures to control water development had environmental effects on Palestinian agriculture. For example, where well drilling for Jewish settlements was carried out in close proximity to Palestinian springs or wells, there often was a marked decline in the output of the springs and a lowering of the water level in the wells.[26] With superior technology in hand, Mekorot drilled deep wells after extensive geological surveying, in contrast to Palestinian farmers who drilled shallower wells in convenient locations.[27] The deeper the well and the more geologically sound its location, the more abundant its water supply and the better equipped it is to resist contamination, salt water encroachment, and the harsh effects of drought.[28] Moreover, when two wells are located within the effective radius of each other, the deeper one tends to milk the water supply of the shallower one. When this is coupled with absence or sparseness of rainfall, the shallower well is gradually sucked dry.

V. Arab and Israeli Usage of West Bank Water

While Palestinian Arabs were prevented from sinking new wells for agriculture, Mekorot continued to drill wells on the West Bank for the domestic and irrigation needs of Jewish settlements. Throughout the 1980s and into the 1990s, the settler population of less than 100,000 has been consuming between one-third and one-half the amount of groundwater consumed by the Palestinian population of roughly 1,000,000.[29] Over time, the inequality of water distribution sharpened. In 1987, *Le Monde* reported that West Bank Arabs were receiving barely more than 20 percent of the total volume of pumped water.[30] That figure dropped to 17 percent in 1994.

Approximately one-third the total land area of the West Bank is cultivated. In 1991, the cultivated area of Jewish settlements represented less than 5 percent (9,030 hectares) of the total cultivated area. However, as much as 69 percent of the Jewish cultivated area is irrigated, as opposed to only 5 percent of the Arab cultivated area.[31] Another striking differential in water use can be found in the Jordan Valley: there, Jewish settlers who farm one-quarter of the cultivated area use 43 percent (40 mcm) of the water consumed in

Table 9.4 Consumption of West Bank groundwater (1993/94)

CONSUMER	POPULATION	CONSUMED
Israelis (inside Green Line)	4,800,000	475 mcm/yr
Settlers (excluding East Jerusalem)	70,000	40–50
West Bank Arabs	1,000,000	105

Source: Compiled by author from *Associates for Middle East Research* Data Base; Lonergan and Brooks 1994.

the valley by agriculture. In contrast, Arab farmers with three times as much cultivated land consume only slightly more water (57 percent or 54 mcm).[32] Furthermore, Jewish settlers in the West Bank consume four times more water per capita than do the Palestinians: about 368 litres per capita per day, as opposed to roughly 88 litres.[33] Although figures vary, the absolute limit on West Bank water use— that is, the sustainable annual yield minus what is lost to evaporation or surface runoff—is about 650 mcm per year. Nonetheless, in the spring of 1990, the combined water allocations for West Bank Arabs, Jewish settlers and the amount reaching Israel was estimated at 807 mcm per year.[34] This represents an annual water deficit in the territory of about 200 mcm, due to the overexploitation, especially in times of drought, of the western and northeastern basins. Overutilization had led, by 1982, to a drop in the water table at a rate of 0.3 to 0.4 metres per annum in the western basin and, according to some sources, of almost two metres in part of the northeastern basin.[35] As indicated above, this constitutes a grave potential danger to the continued intensive utilization of groundwater: a situation that can be reversed only by the systematic recharging of aquifers.

VI. Assessing Solutions:
Lessons from the Jordan Basin

The development and management of river basins as indivisible units, irrespective of political boundaries, has been widely advocated by engineers, planners, politicians, and jurists, and has been adopted as policy in multinational agreements in several basins.[36] Unitary basinwide development is consistent with a functionalist approach to conflict resolution: the idea that ongoing functional cooperation among adversaries and the creation of supranational task-related organizations is the most promising avenue to achieving peace.[37]

On two occasions in the history of the conflict over the Jordan waters, the United States' government responded to the acute need for resource development in the states of the Jordan-Yarmouk watershed by actively supporting functional, basinwide projects. On both occasions, the government hoped that the implementation of these projects would serve as catalysts to peace in the region.[38] As suggested by functionalist theory, regional development was viewed as a stepping stone to regional peace, given that projects would require ongoing multilateral cooperation in the use of water resources. The Eisenhower administration and, later, the Carter administration hoped that solving water problems would provide the climate necessary for resolving the larger interstate conflicts. However, both efforts fell short of their objectives.

On the first of these two occasions, in 1953, a broad, regional plan was drawn up by an American engineering firm to use the water of the Jordan River basin for the irrigation and hydroelectrical power needs of the four riparian states. President Eisenhower appointed Eric Johnston as his personal representative to secure agreement from the riparians to the *Unified Plan*. The mediation effort was conducted in four rounds of negotiation over a two-year period.

Throughout the negotiating process, there was constant disagreement over allocations of water and locations of storage sites, as well as conflicting views of rights, needs, and international legal precedents. The Arab states were concerned about becoming dependent on Israel and having to rely on its goodwill to gain access to water. Israel was chiefly concerned about international supervision of the water distribution and management scheme and the resultant constraints on its national sovereignty. Eventually, many of the contentious issues were resolved, and both sides agreed that as a hydro-engineering plan, the project was acceptable. However, after lengthy deliberations, the Arab League made it known that within the prevailing political context, the project could not be approved; there could be no collaboration with Israel in sharing and managing water resources in the absence of a political settlement of the Arab-Israeli conflict. In 1955, the mediation effort and the Unified Development Plan were put to rest.

In the late 1970s, the United States Government again tried to promote regional water development as a stepping stone to regional peace. This time, the issue was the impounding of the Yarmouk River waters behind a dam at Maqarin, so that Jordan could utilize the winter flow. The Carter administration announced that it would fund a large portion of the project, if Jordan reached an understanding on water allocations with Syria, the upstream riparian, and with Israel, the downstream riparian.

Between 1976 and 1981, the Maqarin Dam project was on the negotiating table, helped along for some of that period by the shuttle diplomacy of then United States Assistant Secretary of State Philip Habib. Again, the negotiations were frought with difficulties. Not only did Israel insist on a much larger allocation of Yarmouk water than what Jordan was offering, but it also claimed the right to take

Yarmouk water to the West Bank—undoubtedly, to supply Jewish settlements.

Because the dam was perceived to be vital to the Hashemite Kingdom's continued survival, Jordan may have conceded eventually to some modified version of Israel's demands. However, the Syrian government would not respond to any overtures regarding a trilateral water-sharing plan. Syria would not agree to a project that entailed any form of cooperation with Israel, nor would it engage in a scheme that would be located within reach of Israeli artillery. The Syrian leadership was acutely aware, from past experience, that hydraulic installations were tempting military targets in conflict situations.[39] No doubt, Hafez al-Asad was also profoundly concerned about the sophisticated installations on the Euphrates River in northern Syria. Thus, much to the chagrin of the Jordanians, this project was abandoned, as well.[40]

The experience with the Maqarin Dam project as with the earlier Unified Development Plan elucidates the proposition that regional political conflict hinders the resolution of water disputes in international river basins. Both episodes suggest that states that are antagonists in the "high politics" of war and diplomacy are not likely to agree willingly to extensive collaboration in the sphere of "low politics," centered around economic and welfare issues. To the adversaries, technical collaboration may be viewed as a disavowal of those issues that fuel the conflict. Hence, an end to riparian dispute requires the prior resolution of political conflict, or considerable progress in that direction. This finding refutes the position and efforts of previous U.S. Administrations. Regional development of water resources could not be realized precisely because of the protracted conflict.

Indeed, along with the status of Jerusalem, water sharing in the Jordan-Yarmouk basin is one of the thorniest issues to be resolved in the current peace negotiations. As one Israeli analyst has noted: "the Arab population in Judea and Samaria may have a claim to the waters that flow underground, and that is one-third of Israel's water supply."[41]

VII. Water and the Middle East Peace Process: 1994–96 and Beyond

On 26 October 1994, Israel and Jordan signed a peace treaty. Not only did this end the state of war between the two countries, it also opened up an era of cooperation in a number of areas, including water resources. In fact, the treaty lays out the terms of a water agreement.[42] By January 1996, progress had been made toward the implementation of at least some of the terms. For one, several joint committees have been created and meet regularly. Moreover, the pipeline that, by the terms of the treaty, was to transport 10 mcm of water per annum from Lake Tiberias in Israel to Jordan, is already in place and functioning.

Compared to the situation between Israel and Jordan, prospects for a mutually satisfactory cooperative solution to West Bank water issues are still uncertain, given that such a large proportion of the water consumed in Israel derives from groundwater that originates in the West Bank and that consumption of this source predates the Israeli occupation of the territory. Equally significant is the fact that the three subterranean basins represent the most important fresh water supply in the West Bank, and thus they would be essential for the future socioeconomic development of a Palestinian state.

While joint Israeli-Palestinian technical committees were set up in keeping with the stipulations of the *Declaration of Principles* (1993), the water resources committee, for example, had made little headway by the spring of 1995. This may have been due in part to problems at the political level and "on the ground": progress toward the transfer of rule in the West Bank to the Palestinian Authority had reached an impasse, and the Palestinian people were feeling uncertain about their future and disillusioned with the peace process.

Then in June 1995, the water resources working group that had been formed within the framework of the multilateral track of the peace process, met for the seventh time. At that meeting in Amman, a number of specific joint projects were discussed and funds were committed by various parties.[43] No doubt, substantive progress was inspired by the negotiations toward the *Israeli-Palestinian Interim Agreement on the West Bank and Gaza Strip* ("Oslo II").

Oslo II was signed by Yitzhak Rabin and Yasir Arafat in Washington, D.C., on 28 September 1995. The agreement spells out arrangements for both transferring authority and managing the shared interests and concerns of the Israelis and Palestinians in the territories until final status negotiations take place. A number of terms of the water component of that agreement deserve special mention. First, while specifying water rights has been put off until final status negotiations, Israel explicitly acknowledges in writing that the Palestinian people do have rights to West Bank water resources. Second, concurrent with the transfer of authority in civil matters to the Palestinians, powers and responsibilities in the sphere of water, and concerning only the Palestinians, will be transferred as well. A permanent Joint Water Committee (JWC) will be established for the interim period to oversee both the implementation of the terms of the Agreement and all matters related to coordinated management, distribution, and exchange of information. Under the auspices of the JWC, a joint mechanism for supervising and enforcing the agreement—Joint Supervision and Enforcement Teams (JSET) —will be created. (It is important to note that by January 1996, both the JWC and the JSET were functioning entities.) Third, while allocations of West Bank water are to remain more or less the same as they were just prior to September 1995, both sides agreed that the Palestinian population of the West Bank would require an additional 70–80 mcm of water per year in the future to meet its needs. During the interim period, both sides would work toward making available an additional 28.6 mcm and developing unutilized sources from the eastern aquifer for the Palestinians.[44]

If progress is made in the bilateral negotiations and peace treaties are signed between each set of adversaries, unitary basinwide development of water resources under supranational authority could eventually be reconsidered. Indeed, in terms of geography, economic efficiency, and human welfare, this represents the ideal solution to the satisfaction of competing needs and conflicting interests in an international river basin. Moreover, if it became apparent that the waters of the basin were insufficient for meeting the needs of the basin states—as in the Israeli-Jordanian-Palestinian arena—then, with a favorable political climate, it may be feasible to "import" water from outside. In recent years, the interest in this solution to water scarcity has inspired a variety of imaginative proposals.[45]

The resolution of protracted conflict would open up vast possibilities for functional cooperation. However, this must not be viewed as the entire solution to resource scarcity in adversarial settings. For, in addition to resolving their political disputes, the states of the central Middle East must take bold measures to curtail excessive and wasteful consumption of water, especially by their agricultural sectors.[46] Concurrently, they must adopt the most effective water-conservation technologies. No doubt, if such technologies were implemented cooperatively, they could have a far greater mutual benefit.

Revamping agriculture does not require the end of regional political conflict. However, if conflict is not brought to an end, states in arid regions will be forced to continue to rely on, at best, suboptimal solutions of a purely domestic and piecemeal nature. At worst, they may be tempted to adopt policies of an aggressive and imperialistic nature in response to the constraints posed by scarcity. Given the stresses on water supplies in Israel, Jordan, the West Bank, and Gaza Strip—among them, the absence of additional, unexploited sources, population growth trends, and recurrent drought conditions—basinwide, and ultimately regionwide, arrangements for sharing, utilizing, and managing water are crucial for human welfare and long-term stability. History shows that a political settlement of the Arab-Israeli conflict, including a mutually acceptable outcome of the final status negotiations, is the essential first step.

--------------------------------- NOTES ---------------------------------

An early version of this paper appeared in *International Security* vol. 18, no. 1 (Summer 1993). The author wishes to thank the journal editors for granting permission to reprint portions of that article. This version was completed in January 1996 and hence does not consider developments in the region since the Likud government of Binyamin Netanyahu came to power in Israel.

1. Information concerning the meetings of the water resources working group has been provided to this author by several participants at those meetings, who will remain anonymous I hereby thank them for their assistance and generosity.

2. Miriam R. Lowi, *Water and Power: the Politics of a Scarce Resource in the Jordan River Basin* (Cambridge, U.K.: Cambridge University Press, 1993; updated 1995).

3. Thomas Naff, "The Jordan Basin: Political, Economic, and Institutional Issues," Guy LeMoigne, Shawki Barghouti, Gershon Feder, Lisa Garbus and Mei Xie, eds., *Country Experiences with Water Resources Management: Economic, Institutional, Technological and Environmental Issues*, World Bank Technical Paper No. 175 (Washington, D.C.: World Bank, 1992).

4. Groundwater refers to that part of rainfall that seeps into the ground and, upon reaching the water table, moves as subterranean flow. Much of the groundwater appears as the perennial flow in springs and wadis, or seasonal streams. Where the geological formations are favorable, groundwater may be obtained as a source of supply by pumping from wells. In order to be considered a gain in total water resources, the supply from groundwater thus obtained must be so located that it would not be recoverable from the springs and wadis.

5.	POPULATION	IRRIGATED AREA (HECTARES)	WATER CONSUMP. (MCM/YR)
1949	1,059,000	30,000	230
1977	3,575,000	186,500	1660
1989	4,200,000	217,000	1950

(Galnoor, 1980, 289; Schwarz, 1992, 130)

6. *Jerusalem Post*, March 24, 1986, p. 1.

7. Itzhak Galnoor, "Water Policymaking in Israel," Hillel I. Shuval, ed., *Water Quality Management under Conditions of Scarcity* (New York: Academic Press, 1980a).

8. Interview by author with a spokesman for Mekorot—the country's National Water Authority, Tel-Aviv, June 15, 1986; Fred Pearce, "Wells of Conflict on the West Bank," *New Scientist*, June 1, 1991, p. 37. See below, section V.

9. Declassified document on West Bank water, provided to author by the Department of State, Government of the United States (n.d.). Hereafter cited as Dept. of State.

10. Jehoshua Schwarz, "Water Resources in Judea, Samaria, and the Gaza Strip," Daniel Elazar, ed., *Judea, Samaria, and Gaza: Views on the Present and the Future* (Washington, D.C.: American Enterprise Institute for Public Policy Research, 1982).

11. Ibrahim Dakkak, "Water Policy on the Occupied West Bank" [al-Siyaasa al Ma'iya f'il Difa al-Gharbiya al-Muhtala], *Shu'un Filastiniyah* vol. 126 (May 1982); Kahan, David, *Agriculture and Water in the West Bank and Gaza* (Jerusalem: West Bank Data Base Project, 1983); Shawkat Mahmoud, "al-Zira'a' wal-Miyaa f-il-Dafa al-Gharbiya taht al-Ihtilaal al-Isra'ili" (Agriculture and Water on the West Bank under the Israeli Occupation), *Samed al-Iqtisadi* vol. 6, no. 52 (November–December 1984); Thomas Naff, "The Jordan Basin: Political, Economic, and Institutional Issues," Guy LeMoigne, Shawki Barghouti, Gershon Feder, Lisa Garbus, and Mei Xie, eds., *Country Experiences with Water Resources Management: Economic, Institutional, Technological and Environmental Issues*, World Bank Technical Paper No. 175 (Washington, D.C.: World Bank, 1992); and Jehoshua Schwarz, "Water Resources in Judea, Samaria, and the Gaza Strip," Daniel Elazar, ed., *Judea, Samaria, and Gaza: Views on the Present and the Future* (Washington, D.C.: American Enterprise Institute for Public Policy Research, 1982).

12. "The Zionist Organization's Memorandum to the Supreme Council at

the Peace Conference," in J. Hurewitz, *Diplomacy in the Near and Middle East*, vol. 2 (Princeton, N.J.: D. Van Nostrand, 1956), p. 48.

13. H. F. Frischwasser-Ra'anan, *The Frontiers of a Nation* (Westport, Conn.: Hyperion Press, 1955).

14. Walter Laqueur, *A History of Zionism* (New York: Schocken Books, 1972).

15. Interview by author with an Israeli geomorphologist, Haifa, June 21, 1986.

16. Quoted in Michael Brecher, *Decisions in Israel's Foreign Policy* (New Haven, Conn.: Yale University Press, 1975), p. 184, from *Divrei Ha-Knesset*, vol. 15, pp. 270–71, November 30, 1953.

17. Michel Foucher, "Israel-Palestine: quelles frontières?" *Hérodote*, 2è–3è trimestres (1983) Dan Horowitz, "The Israeli Concept of National Security and the Prospects of Peace in the Middle East," Gabriel Sheffer, ed., *Dynamics of a Conflict: A Re-examination of the Arab-Israeli Conflict* (Highlands, N.J.: Humanities Press International, 1975); Michael Mandelbaum, *The Fate of Nations: The Search for National Security in the 19th and 20th Centuries*, (Cambridge, U.K.: Cambridge University Press, 1988).

18. Uri Davis, Antonia Maks, and John Richardson, "Israel's Water Policies," *Journal of Palestine Studies* vol. 9, no. 2 (Winter, 1980); and Jeffrey D. Dillman, "Water Rights in the Occupied Territories," *Journal of Palestine Studies* vol. 19, no. 1 (Autumn, 1989).

19. Meron Benvenisti, *The West Bank Data Project: a Survey of Israel's Policies* (Washington, D.C.: American Enterprise Institute for Public Policy Research and West Bank Data Base Project, 1984); Jeffrey D. Dillman, "Water Rights in the Occupied Territories," *Journal of Palestine Studies* vol. 19, no. 1 (Autumn, 1989); and Dept. of State.

20. Kahan, David, *Agriculture and Water in the West Bank and Gaza* (Jerusalem: West Bank Data Base Project, 1983).

21. Interview by author with Israeli water engineer, formerly with Tahal, Tel-Aviv, June 26, 1986.

22. House Committee on Foreign Affairs, *Middle East in the 1990s*, p. 187. This is a curious regulation. In hot, dry climates, the best times of day for irrigating are early in the morning and late in the afternoon or evening. Irrigating at night is not uncommon. At these times, there is significantly less evaporation than there is between 10 A.M. and 4 P.M.—the hottest time of day. Moreover, there is less risk of destroying plants and their roots. Intense watering turns soil to mud; and when hot sun beats down on mud, plants rot.

23. Interview by author with former Deputy Mayor of Jerusalem, 8 June 1986; Dakkak 1982; Kahan 1983; Mahmoud 1984. The total quantity of water available to West Bank Palestinians for agricultural, municipal, and industrial consumption is about 125 mcm per annum. Of the approximately 100 mcm of water consumed in agriculture, only 38 mcm is pumped directly from wells; the remainder derives from springs.

24. House Committee on Foreign Affairs, *Middle East in the 1990s*, p. 187.

25. Thomas Naff, in his testimony before the House Committee on Foreign Affairs, *Middle East in the 1990s*.

26. Dakkak 1982; Paul Quiring, "Israeli Settlements and Palestinian Rights, Part 2," *Middle East International*, October 1978, pp. 14–15; Khalil Touma, "Bethlehem Plan Further Threatens Scarce Water Resources," *Al-Fajr*, 26 July 1987, pp. 8–9; "Proposed Israeli Well Will Deepen West Bank Water Crisis," *Al-Fajr*, 5 July 1987, p. 3; "Water Supply under Occupation," *Al-Awdah* English Weekly, 12 October 1987, pp. 19–23; United Nations, Committee on the Exercise of the Inalienable Rights of the Palestinian People, "Israel's Policy on the West Bank Water Resources" (New York: U.N., 1980), pp. 13–15.

27. Dept. of State, op. cit.

28. Unlike Arab wells, which rarely exceed depths of 100 metres, those drilled by Mekorot are between 200 and 750 metres deep (Kahan 1983; Mahmoud 1984). As noted above, greater depth allows for superior output and water quality.

29. Ibrahim Dakkak, "Water Policy on the Occupied West Bank," op. cit.; Dept. of State, op. cit.; David Kahan, *Agriculture and Water in the West Bank and Gaza*, op. cit.; and Stephen Lonergan and David Brooks, *Watershed: The Role of Fresh Water in the Israeli-Palestinian Conflict*, op. cit.

30. J.-P. Langellier, "Guerre de l'Eau en Cisjordanie," *Le Monde*, November 7, 1987.

31. H. Zarour, and J. Isaac, "The Water Crisis in the Occupied Territories," paper presented at the Seventh World Congress on Water Resources, Rabat, Morocco, 12–16 May 1991; and H. Zarour, and J. Isaac, "Nature's Apportionment and the Open Market: A Promising Solution to the Arab-Israeli Water Conflict," *Water International* vol. 18, no. 1 (1993), pp. 40–53.

32. For confirmation of these figures, see Schedule 10 of the *Israeli-Palestinian Interim Agreement on the West Bank and Gaza Strip* (September 1995).

33. Kolars (1990) and an Israeli geographer interviewed in 1993 gave figures of 300 litres and 76 litres on average. Note that a per capita consumption of 100 litres of water per day is generally considered to be the minimum for an acceptable quality of life.

34. John Kolars, "The Course of Water in the Arab Middle East," *American-Arab Affairs* no. 33 (Summer 1990).

35. The (south)eastern water table, the smallest of the three West Bank aquifers, is the principal Palestinian water source and the only one that is not yet overused. According to the Oslo II agreement (see below, section VII), the aquifer contains 78 mcm of water that remain unutilized. Jehoshua Schwarz, "Water Resources in Judea, Samaria, and the Gaza Strip," Daniel Elazar, ed., *Judea, Samaria, and Gaza: Views on the Present and the Future* (Washington, D.C.: American Enterprise Institute for Public Policy Research, 1982).

36. C. B. Bourne, "The Development of International Water Resources: The Drainage Basin Approach," *The Canadian Bar Review* vol. 47, no. 1 (March 1969), pp. 62–82; A. H. Garretson, et al., eds., *The Law of International Drainage Basins* (Dobbs Ferry: Oceana Publications, 1967); Albert Lepawsky, "International Development of River Resources," op. cit.; Ludwick A. Teclaff, *The River Basin in History and Law* (The Hague: Martinus Nijhoff, 1967).

37. David Mitrany, *A Working Peace System* (Chicago: Quadrangle Books, 1986 [first published 1943]).

38. Lowi, Miriam R., *Water and Power: The Politics of a Scarce Resource in the Jordan River Basin*, op. cit.

39. During the mid 1960s, the preliminary work in Syria on the Arab scheme to divert the headwaters of the Jordan system succumbed to Israeli gunfire. In June 1967, the construction of the *Mukheiba* dam on the Yarmouk River was stopped by Israeli military intervention. Then, on four separate occasions between 1967 and 1971, Jordan's East Ghor Canal was knocked out of service by Israeli forces. See Lowi (1993/95), chapter 5.

40. The project was temporarily revived in 1987 in the form of a considerably smaller Dam at a different site on the river. Jordan and Syria signed a treaty to build the Unity (al-Wahdah) dam, and Jordan petitioned the World Bank to finance the project. Once again, assistance was conditional upon reaching basinwide agreement. To this end, U.S. Secretary of State James Baker appointed Richard Armitrage as the U.S. mediator between Israel and Jordan. (The U.S. government refused to include Syria in the talks at the time because of the poor relations it had with that regime.) Bilateral negotiations

continued until August 1990, when Iraq invaded Kuwait. The talks were not reconvened following the Gulf War because of the inception of the Middle East peace process. (Personal interviews with U.S. government officials, Washington, D.C., April 1991, 1993.)

41. Interview by author with an Israeli geographer and water resources specialist, Haifa, 22 June 1986.

42. The water agreement, Annex II of the *Israel-Jordan Treaty of Peace*, is reprinted in Appendix 5 of Lowi, *Water and Power*, op. cit., pp. 212–17. For an analysis of what motivated Jordan to sign a treaty with Israel, see Lowi, "Rivers of Conflict, Rivers of Peace," *Journal of International Affairs* vol. 49, no. 1 (Summer 1995), pp. 140–41.

43. Perhaps most important among them is the Israeli-Jordanian-Palestinian regional data banks project to gather, analyze, process, and archive data in such a way that it can be shared meaningfully at some future date.

45. It is interesting to note that, according to Schedule 10 of the agreement, the utilization of West Bank water in September 1995 was gauged at 118 mcm (19.5% of the total) for the Palestinians and 483 mcm for the Israelis.

45. In the early 1980s, for example, Jordan considered the feasibility of a project to carry about 160 mcm of water through a pipeline from the Euphrates River in Iraq to the kingdom's northern plateau (Taubenblatt 1988). A Tel-Aviv University–Armand Hammer Fund project drew up a Middle East water-sharing plan that includes details for the transfer of Nile water to Israel, Jordan, the West Bank, and Gaza Strip, and the transfer of Litani water to Israel, Jordan, and the West Bank (Kally 1986). Turkey's late president Turgut Ozal put forward the idea of a "peace pipeline," which would transport water from two Turkish rivers southward to Syria, Jordan, Saudi Arabia, and the other Gulf states (Duna 1988; Kolars 1990). And in recent years, a Canadian company has been trying to market a scheme to transfer as much as 250 mcm of water from Turkey to Israel in enormous "Medusa" bags, floated accross the Mediterranean (Lonergan & Brooks 1994).

46. This means that they must reconsider the size and importance of agriculture in their economies, revise the choice of crops grown, and phase out the production of water-intensive crops.

10

Imminent Political Conflicts Arising from China's Population Crisis

Jack A. Goldstone

One of the key aspects of the physical environment is the availability of agricultural land. In the modern world, the amount and quality of that land is not necessarily a constraint on economic development or incomes; witness the prodigious progress of Japan and South Korea. Yet for a country whose population is largely composed of peasants who depend on the land, shortages of arable land can have important consequences for patterns of employment, migration, and inequality. As arable resources become scarcer, and require more capital to boost output, subsistence and small-scale peasant farming becomes both less viable and less attractive for many. Inevitably, large-scale migrations from the land to cities or rural industrial enclaves become necessary to sustain growth.

While rapid economic growth can eventually absorb such changes in employment, the period of transition is often a difficult period for political institutions. The shortage of rural agricultural employment and the rapid growth of urban and industrial centers often lead to large increases in inequality; large-scale migration raises the demands for new services and increases the difficulty of governing a highly mobile population; and the rapid growth of indus-try can lead to the formation of new interest groups and new demands for the redistribution of economic and political power. If political institutions lack strong legitimacy, and if elites are factionally divided, these strains can lead to open conflict and the possible breakdown of governing institutions. Such a pattern is beginning to develop in the world's most populous agrarian country, China.

For its first three decades, Communist China consciously disregarded the impending clash between its massive population and its limited arable and environmental resources. Guided by Mao Zedong's beliefs that every worker came equipped with his labor to sustain him (or her), and that only the capitalist exploitation of such labor for profits stood in the way of rising living standards, the Party positively encouraged both population growth and heedless use of land, water, and mineral resources, confident that socialism would somehow overcome the physical constraints of supporting an ever larger population on an ever more strained and degraded stock of natural resources.

By Mao's death in 1976, the failure of socialism to overcome these problems was evident. The current Party leadership thus finds itself in a dilemma: population has grown enormously; land, water, and forest resources have been eroded, polluted, decimated, and cannot provide adequate employment for China's still largely peasant population; and socialism's failure to deliver China from this impasse has discredited it as a guiding ideology among many Chinese. These daunting problems threaten the Communist Party's identity and legitimacy as the vanguard party of socialism. Deep divisions and conflicts in China have arisen in the last fifteen years as the Party leadership has tried to find an alternative economic policy to solve the looming resource crisis while preserving its claim to socialist leadership.

China's leaders have come to recognize that population growth poses a threat to China's political stability and economic growth. Their consequent implementation of a one-child family policy was intended to avert the worst. However, even draconian measures cannot suddenly halt the demographic momentum of a large, relatively youthful, population. The large cohorts of children born in the late 1960s and 1970s—when Mao encouraged large families to strengthen China—are now entering their prime child-bearing years. Even with China's one-child policy in place, its population will increase by 25 percent, or almost three hundred million people, in the next twenty years. Given recent relaxations and evasions of the one-child policy, especially in the countryside, where 80 percent of China's population remains, the increase could well be greater.[1]

The ability of China's natural resource base to provide a livelihood for those hundreds of millions has limits. Although coal and potential hydroelectric resources may provide adequate power, China is deficient in the most basic commodity for an agricultural, overwhelmingly peasant, society: arable land. Not only is arable land scarce, it is decreasing, due to erosion and expansion of urban settlements and transportation networks. In the next two decades, arable land may diminish by 10 percent; given the impending population increase, this will result in a decline in arable land per capita of just over 25 percent.

To make matters worse, the productivity of existing land will not easily be increased to make up the difference. The quality of the land is being reduced by salination from heavy irrigation and use of artificial fertilizers, by erosion of topsoil, and by reduced efficiency of irrigation due to siltation of rivers and irrigation channels. Enormous capital investments will be required in order to maintain the high productivity of China's farms and fields.

In addition to these strains on farmland, woodland has been subject to equal or greater degradation, due to deforestation and erosion of forested slopes, and to large-scale illegal logging to provide timber for everything from fuel and housing to mine-shaft supports. Air and water quality have been diminished by industrial effluents. This degradation of the arable farming environment has provoked migration to China's cities, where vast populations and lagging public services have aggravated pollution of the urban environment.

Despite the novelties imposed by industrialization, the Communist Party-state, and modern communications technology, the overall pattern described above—a growing environmental imbalance between increased population and overburdened arable land—has developed several times in China's long imperial history. Each time, the result has been a constellation of political conflicts: between regions, between the capital and the provinces, and between different elite and popular groups. In the seventeenth, nineteenth, and twentieth centuries, these conflicts produced regional rebellions, massive civil wars, and the collapse of the central state.

In order to assess the threat of similar conflicts arising in the future, I first briefly survey the conjunction of environmental stresses and political crises in Chinese history. I then examine the political vulnerabilities of the current Communist Party-led state. In conclusion, I examine the difficult choices imposed by the current environmental crisis on the Communist state, and demonstrate that any conceivable choices will lead to severe conflicts, and the probable dissolution of Communist rule, with attendant uncertainties and chaos.

Environmental Crises and Political Crises in Chinese History

From the beginning of the Ming dynasty in 1368 to the end of the Manchu dynasty in 1911, China's political stability has been tied to the balance of population and land.

As a result of steady population growth during Ming rule, the amount of cultivated land per capita is estimated to have fallen by 33 to 50 percent from the early fifteenth to the mid-seventeenth century.[2] By the early seventeenth century, the imperial treasury was having difficulty collecting taxes from peasants whose growing families overburdened their declining land holdings. Rebels, recruited from unpaid deserting soldiers and desperate peasants, captured Beijing in 1644, destroying Ming rule and paving the way for better organized Manchu armies from the northern steppes to begin their conquest of China. Separate rebellions carved out a new regional power based in Sichuan/Hunan, and new "Ming loyalist" states appeared in the south. It took several decades for the Manchu to subdue these varied regional powers and reunify China.

The late Ming rebellions and the Manchu conquest took a fearful toll of lives, but this loss restored the balance between population and land. After 1660 in "Szechuan, Yunnan, Kwiechow, Shensi, and Kangsu . . . underpopulation and abundance of land were manifest."[3] Rulers of the new Manchu (Qing) dynasty further expanded the area of the empire, adding the lands of Manchuria to the north and central Asia to the west, and sponsored extensive land reclamation and resettlement programs. As a result, land cultivated per capita appears to have increased by 50 percent between 1600 and 1730, and to have remained ahead of late Ming levels at least until 1770.[4] The early Qing was thus a period of prosperity, and once the Manchu conquest was completed, a period of great domestic stability.

Yet this prosperity did not last, for it was soon overwhelmed by renewed population increase. In Hunan province, as early as 1748 gazetteers complained of rising rents, land shortages, and "the growing pressure of population on a province that was exhausting the supply of readily cultivable land."[5] Toward the end of the eighteenth century, farmers on overcrowded lands began to see a reversal of earlier income gains.[6] By 1850, a disastrous imbalance had again developed: China's population appears to have increased to nearly 400 million inhabitants, half again as many as in 1770; but cultivated acreage had increased by only a quarter. Cultivated land per capita sunk even lower than it had been in the early seventeenth century.[7] Rebellions again were the result: the Qing dynasty faced revolts in the west and northwest, in Hunan, Sichuan, Henan, Zhili, and Shandong. This wave of revolts climaxed in the Taiping Rebellion (1850–1863), which ranged throughout central and southern China, and killed literally tens of millions. The Qing dynasty survived only through the eccentricities of the Taiping leaders and the loyalty of gentry elites who raised private armies to suppress the rebels.

Although tragic, the vast casualties of the Taiping period briefly restored a balance between land and population; but the respite lasted only a few decades. By the end of the nineteenth century, population had recovered its losses, and land scarcity again bore down on the rural population. Amid a variety of revolts by regional warlords, the Qing dynasty fell in 1911. Again, China did not merely fall, it fell apart, as unified rule was replaced by regional warlords. China was not unified under central rule again until the Communists established nationwide rule almost four decades later.

These past episodes of population growth and political conflict reveal several consistent patterns. First, political conflicts were shaped not only by growing poverty, but also by growing prosperity. Population growth meant growing poverty for the relatively arid regions of the northwest—Gansu, Shaanxi, Shanxi. Poverty also increased among the peasants of the densely populated, but geographically tightly bounded, Sichuan plain, and among the land-poor peasants of the lower Yangzi basin. Not surprisingly, these regions—particularly Shaanxi and Sichuan—have been the traditional home of regional rebellions against the central authorities. However, growing prosperity was also brought by population growth, for the wealthy landowners and traders of the lower Yangzi and southern coastal regions—the area ranging from Shanghai through Fujian and south to Guangdong—profited from growing demand for basic commodities. Yet this prosperity also brought conflicts. The newly enriched merchants and landowners of these regions often resented and resisted the authority of the imperial government in Beijing, which sought to divert this commercial wealth to its own treasury. It was this two-pronged attack on the central authorities—elite resistance to taxation that handcuffed the state, combined with regional popular revolts—that overturned dynasties.

Second, past conflicts were exacerbated by migration streams that inflamed regional antagonisms. Conflicts over land between farming migrants and indigenous populations strained the ability of the state to keep order, and undercut economic progress. In addition, migration to the cities tended to lower real wages, providing fuel for urban riots and revolt.

Third, the survival of the ruling regime depended on retaining the loyalty of key elite groups—primarily the army and bureaucratic officials. When army leaders and key bureaucrats remained loyal to the regime, revolts could be put down. Such was the case for the Nien, White Lotus, and Taiping rebellions. However, when army and bureaucratic leaders declined to make efforts to support the regime, and when ordinary soldiers deserted to join peasants, workers, and intellectuals in revolt, as at the end of the Ming and Qing dynasties, the regimes collapsed.

Fourth, population pressures and political conflicts grow and cumulate in tandem. That is, the early phases of population pressure on land tend to produce limited rebellions in particular regions that are put down. However, continued population growth leads to growing political conflicts, and more severe and widespread rebellions. Stability has only ensued when either drastic population decrease or expansion of arable land, or both, has significantly improved the amount of arable land per capita.[8]

Fifth, the dissolution of dynasties in China has generally been marked by regional rebellions and civil war, and resulted in the temporary dissolution of the vast Chinese state into regional warlord domains or independent states.

These historical patterns provide considerable insight into the likely dynamics of current political conflicts in China produced by environmental pressures.

Political Vulnerabilities of the Current Regime

The end of the Qing dynasty did not lead to stability. Population pressures and political conflicts have continued throughout the twentieth century. The warlords who controlled China after 1911 were only briefly brought into line by the Nationalist leadership of Chiang Kai-shek; the Nationalists themselves were overthrown by the Communists under Mao in 1949. Each period of civil war has brought a decade or two or three in which population losses and agricultural expansion combined to create an initial period of prosperity. However, China's population increase has been so relentless and so massive—increasing from four hundred million in 1850 to over one billion in 1980—that no twentieth-century regime has been able to evade its pressures.

Mao Zedong made two efforts to cope with population growth by central command. The first effort consisted of diverting labor into inefficient rural smelting and backyard industries. The second consisted of commanding the planting of grain on every plot of cultivated land, regardless of suitability or cost. Both efforts were unmitigated disasters. The diversion of labor to backyard industries resulted in a famine that cost perhaps thirty million lives in 1959–1961.[9] The attempt to spread grain planting to all cultivated land led to destruction of woodland, massive erosion, and fertility exhaustion of land that had been productive of potato, peanut, or other crops.

The failure of Mao's policies prompted the reforms of Deng Xiaoping, begun in 1979. Deng implemented three major reforms that were quite successful in coping with population growth and increasing prosperity. First, Deng abandoned centrally controlled, communal farming, returning land to private control by peasant families. This enormously increased productivity by encouraging more efficient and diligent use of farmland. Second, Deng encouraged foreign trade and production for profit, setting up new economic zones in Shanghai and the south coast where entrepreneurship in trade and manufacturing could grow. Third, population growth was attacked directly through the one-child policy, which temporarily drastically reduced population increase.

Deng's measures reversed the worst disasters of Mao's rule. Yet these measures will not greatly increase the supply of arable land, nor, given the demographic momentum of China's young population, will they stop population growth until the middle of the next century. Thus we must still assess the political pressures likely to arise from increased strains on China's resources for decades to come.

The current regime has a number of vulnerabilities reminiscent of those that toppled past dynasties in Chinese history. The combination of forces revealed in the Tiananmen Square Uprising of 1989—a coalition of merchants, entrepreneurs, urban workers, students, and intellectuals, with some support from within the regime, in revolt against a government that survived only because of continued loyalty of key military and bureaucratic leaders—is quite similar to that of past patterns of Chinese revolt. Past experience also suggests that as population strains continue, political conflicts will increase.

The first major vulnerability of the regime is the split within the ranks of the Party leadership revealed in spring of 1989. As Andrew Walder observed, the spring uprising "could never have grown to the size it did, nor could it have shaken the halls of power to such an extent, had the party's . . . political control not been weakened over the decade of reform and had there not been open divisions within the party apparatus and leadership itself."[10] This split in the Party leadership is complicated, and appears to run across three issues, rather than a single dividing line.

First, the Party leadership is divided over the amount of central, socialized control to maintain over the economy. The economic reform faction, led by Zhu Rongji, advocates a loosening of socialist control of the economy. The continued success of this faction is shown in the spectacular growth of China's "special economic zones," especially along the southeastern coastal regions, and in the appointment of economic reformers to major posts even after the spring uprising. The enormous success of private entrepreneurs in these special zones, along with the support of overseas Chinese for such liberalization, has made for an increasingly important and vocal constituency for continued reform. However, there are still numerous Party officials, not to mention hundreds of millions of timid workers and peasants, who fear the unrestricted workings of the market. Party leaders and lower-level functionaries fear loss of their power and positions; workers and peasants fear the loss of security provided by permanent state jobs and the attendant risks of the free market. Given this substantial constituency for conservatism, Party leaders who fear change, and lower-level Party officials who seek to obstruct it, continue to limit the ability of the Party to embrace full-fledged national economic reform.

Party leaders also are divided over the extent to which the Party should surrender more democratic political control. Party leader Jiang Zemin, although committed to market reform, would prefer to implement economic changes under strict Party control rather than allowing popular intervention in policy decisions. However, much of the population, particularly the students, professionals, workers, and aspiring entrepreneurs, lost faith in the Communist Party's leadership during the tumultuous years of Mao Zedong's rule and the brief reign of his wife and the "Gang of Four" that followed. A large popular constituency—both economically reformist and economically conservative—as well as a substantial faction of Party leaders, believe that China would fare better under a government more responsive to popular needs and demands. Nonetheless, Jiang believes that democratization at this point in time would undermine the Party's efforts to modernize China's economy. This split among Party officials was even more fundamental in the crisis of spring 1989 than conflicts over economic policy.

If the first two issues of conflict are about goals—how far and fast to liberate the economy from socialist control, and how far and fast to democratize the political system—the third issue is about the means of dealing with the growing conflicts over goals. At the time of the Tiananmen uprising of 1989, the moderate faction, led by now-discredited former Party General Secretary Zhao Ziyang, believed that many popular and elite demands regarding government policy were legitimate, and that solutions should be negotiated in a spirit of compromise between the Party and Chinese society. The hard-line faction, led by then-President Yang Shangkun and Premier Li Peng, believed that the decisions of the Party must be accepted, and that any resistance or opposition to the Party from Chinese society is a dangerous counterrevolutionary force which must be ended, and crushed by force if need be. It was this split, even more than divisions on policy issues, that paralyzed the leadership of the Party during the spring uprising. It was only when Deng Xiaoping threw his weight firmly behind the hard-line faction, and gained sufficient support for this faction among regional army commanders, that the issue of how to deal with the student uprising was settled.

Continuing conflict over these three issues means that consensus on goals and methods within the Party leadership is weak. Deng himself had been an economic reformer, but a political conservative and a hard-liner on methods. Yet other key figures on the Politburo do not generally share this particular set of beliefs. Many of the leading economic reformers in the Party whom Deng has assigned to preside over the new economic policies, such as Zhu Rongji, recognizing their reliance on support for the reforms from entrepreneurs, investors, and workers, would prefer more democratization to accompany the economic reforms. On the other hand, some of the key hard-liners who supported Deng's Party crackdown appear to be economic conservatives, such as former President Yang, Premier Li Peng, and secretary-general Jiang Zemin. The result of these cross-cutting Party divisions is that neither Jiang (nor anyone else) has a secure and unified cadre to carry out their wishes. Instead, each action must be negotiated among different factions, with force as the last resort to sway recalcitrant Party officials. And in a crisis, the lack of unity and consensus means paralysis until one faction can find a way to prevail. In short, the Party leadership at this time is extraordinarily ill-suited to pursue a strong and radical course of economic or political reform, and ill-equipped to deal with political or economic crises.

The second major vulnerability of the regime is the rift with China's broader elites. With the growth of special economic zones, the development of modern communications, and the decline of the military threat from Russia, elites outside the Communist Party civilian leadership—business leaders, army leaders, students, intellectuals, and professionals—are searching for and developing new roles. These roles are often in conflict with the policies of the current Party leadership. For example, students, intellectuals, and many entrepreneurs favor more dialogue and compromises between the Party and non-Party elites. Although they often do not favor real democratization, maintaining the millennial Chinese ethic of rule by moral and intellectual elites, they nonetheless are pushing for more independence from the Party and more rapid democratization than Jiang would like.

These elites were the prime driving force behind the spring 1989 uprising in Beijing and in other urban centers. Unfortunately for China's Party leaders, it is impossible to pursue their policy of modernization, and particularly of economic reform in the special economic zones, without increasing the number and importance of such student, business, and professional elites. Historically, the crisis point of past autocratic regimes—such as Russia in 1917 and 1989, France from 1789 through 1848, and Iran in 1979—came precisely at the time they were creating large numbers of such new professional elites to assist in modernizing their societies, but still resisted efforts to incorporate those elites more fully into political life through democratizing reforms.[11] In this sense it is not, as Marx said, modern capitalist economies, but any autocratic regime that seeks to preserve absolute power while encouraging the growth of modern business and professional elites that is creating its own "grave diggers."

Moreover, as China pursues its modernization policies, it requires investments. Given China's limited resource base, such investments inevitably conflict with investments in the military, which can no longer be justified in terms of external security threats. Military pay scales and privileges have recently been reinforced, but are likely to lag in the future. The Party thus faces another dilemma: if it invests heavily in the multimillion-man army to maintain loyalty and morale, the slowdown in economic investment threatens the loyalty of nonarmy elites and the general population. On the other hand, if it does not pamper the army, divisions and disloyalty are likely to grow. Indeed, given the decline of external threats and the divisions within the highest levels of Party leadership, it is unlikely that the military will again have the clear mission and unqualified support for the Party it had under Mao.[12]

The third major vulnerability is the decline in the Party's direct control of Chinese society. The Chinese Communist Party rarely needed the army in the past to exert control over China's society. Controlling every job in every factory, every allocation of land, fertilizer, and seed in the countryside, and controlling every means of communication inside of China, the Party was able to prevent any opposition from reaching a point where military suppression was needed. Such direct central control was a major achievement of Mao's regime, and remained effective for a society of modest urbanization and state-run collective farms. Yet Deng's reforms, and the continued plans for economic modernization, have meant that "the strict control over travel and information and means of communication characteristic of the Mao era has . . . been loosened, as a consequence of both the policy of openness and of China's rapid development."[13] The spread of private motorcycles and telephones has greatly increased the mobility of people and information not subject to government control. As Walder remarks, "The rapid growth in these means of communication has clearly outpaced the capacity of the authorities to monitor them. Perhaps the most dramatic example of this fact is the ability of some hunted activists to escape from Beijing, or from China altogether, in the weeks after the June 4 massacre."[14] The privatization of land in the countryside, and the privatization of factories and their jobs in the special economic zones, has meant that the Communist Party is no longer the exclusive lord of land and labor in Chinese society. To an increasing degree, therefore, the Party will have to rely on voluntary cooperation, or increasingly dubious direct military repression, rather than direct control to implement its policies and maintain its authority.

The fourth major vulnerability is continued discontent among the foot soldiers of the revolution—peasants and workers. Under Mao, the overwhelming direct control of the Party, the unity of Party, and the support of army and professional elites on most matters of policy and methods, meant that popular discontent with the effects of Communist rule could be disregarded as a temporary problem, to be ended with the economic triumph of socialism; it certainly posed no threat to the Party regime. Neither the famine of the late 1950s, nor the chaos of the Cultural Revolution, led to a popular opposition movement that seriously threatened the Party. Direct control meant that popular forces had no room nor resources with which to form an opposition, and substantial unity within the Party and the lack of autonomy among non-Party elites meant that popular groups lacked leadership. Under the conditions prevailing today, however, popular discontent is a problem of far greater dimensions. Peasants and workers discontent with their trials under socialism will find ready leadership among student, business, and professional elites, and even legitimation from within the ranks of Party leaders, for their complaints.

The ups and downs of popular living standards in China under Communism are controversial. During certain periods, particularly the first decade of the Deng reform era, gains are undeniable. On the other hand, for the Communist era as a whole, periods of substantial gains appear to have alternated with periods of inflation, crises, and recession. Perhaps all that is certain at this point is that the next two decades will see a population explosion that will place all the gains of the last three decades in jeopardy. If economic growth is to be sufficient to sustain, let alone improve, living conditions in China for its larger population of the next century, such growth will have to come from industrialization and the growth of manufacturing. There are simply not the resources of arable land or forest to provide for rising living standards for growing populations through expanding the agricultural sector. The route to modern industrial society is well-documented, and it involves moving the population out of agriculture into industrial, chiefly urban, livelihoods. This change has just begun to gain momentum in China, and it must accelerate if China is to avoid destitution.

Nonetheless, this change cannot be wholly welcomed by China's leadership. A massive urban migration will have immense political costs. To absorb such a large migrant population in currently stagnant, money-losing state sector factories will bankrupt the state; but to absorb such a large population in the private sector will so further weaken Party direct control of manufacturing, and so increase the volume and power of non-Party elites as to make their demands overwhelming. It is therefore most likely that Party leaders will continue their current policy of trying to manage and limit urban migration and expansion of the private sector. However, this policy of periodically retrenching the state sector, slowing urban migration, and controlling the private sector, will again lead, as it has in the last decade, to recession, housing shortages, and heavy burdens on farming communities. In short, the attempt to "manage" this migration, in combination with the massive economic strains coming from the population growth in rural China, will likely sustain popular discontent with Party control, for both workers and peasants, precisely at the time that other trends and results of modernization policies are increasing the vulnerability of the regime to such discontent.

In sum, the problem for China's regime is not that popular discontent is likely to grow much greater than it was under the previous periods of hardship in the Mao era. It is not. However, discontent is unlikely to go away, given the looming population/resource crisis, and the regime is becoming far, far more vulnerable to popular discontent than was the case in previous decades. Given the rise in the ability of the population to move, mobilize, and express its discontent, and the growth of an elite opposition among the broader Chinese elites, it would be imperative for the Party to gain a large increase in popular support in the coming decades. This appears, however, a virtual impossibility, as long as the Party clings to full control and responsibility for what will be an increasingly difficult economic situation in the near future.

The various vulnerabilities listed above—a divided Party, opposition from broader elites, the decline of direct control, and a discontented, increasingly mobile population—would not in themselves threaten to overturn the regime. However, when combined with the impending growth of population, and the legacy of three decades of contemptuous neglect of the problem of growing population/resource imbalances under Mao, these vulnerabilities assume great significance.

China's population/resource situation is reaching a point where dramatic action will be required to fend off a major economic crisis. Yet given the regime's vulnerabilities, as noted above, the Party is ill-equipped either to quickly and smoothly implement dramatic action or to cope with a major crisis. As demographic and physical constraints are making this choice unavoidable, the regime will spend the coming decade pinned into a corner from which it is unlikely to escape intact.

Economic Constraints and Political Chaos

China's leaders face difficult choices in the years to come. The rapidly growing population of the north and west cannot be fed and employed within those regions; there is not sufficient land, nor sufficient water, to provide for the additional hundreds of millions that will be born in the next decades. Tens of millions must therefore move—to the farmlands of the better-watered south, and to the cities—to find a livelihood.

If the authorities attempt to block this migration, they will undoubtedly provoke subsistence-based revolts from the desperate populations of the northwest. If these revolts combine with resistance from the merchant-entrepreneurs and workers of the major cities, anxious to remove the burden of Communist supervision, the regime is likely to fall. Indeed, the absence of such regional popular revolts is precisely what allowed the regime to suppress the Tiananmen Square and other urban uprisings in June 1989.

If the authorities direct this migration to the southern countryside, they will buy only a short time until the combination of migration and indigenous population growth with soil erosion and woodland degradation lead to subsistence-based revolts in the south. In this case, disorders will likely be exacerbated by local and regional conflicts between migrants and indigenous farmers.

If the authorities direct this migration to the cities, they will vastly increase those groups over which the regime has the least control—entrepreneurs, urban professionals, workers in private business, and the free-floating unemployed and underemployed. This strategy, of expanding the special economic zones, facilitating urban expansion and the growth of private industry, and encouraging foreign trade to gain food from grain-surplus nations in exchange for China's manufactures, will inevitably have to be implemented; otherwise the Party's rule is likely to be overthrown by regional revolts protesting drastic declines in people's livelihood. This much is recognized by most Party officials, and this strategy was officially ratified at the fourteenth National Party Congress held in October 1992. Yet while the strategy of urban expansion is the only route that promises to provide a livelihood for China's new tens of millions, it also leads directly to the diminished control of the Communist Party over Chinese society.

It is true that Brazil, Mexico, Chile, South Korea, and Taiwan have displayed impressive economic growth through private investment under authoritarian regimes. However, in all these cases the regime was a military or party regime that was internally unified, acted in partnership with private capitalists, and was supported by extensive outside capital, whether coming from the World Bank, commercial bank loans, U.S. A.I.D., or other sources. Once foreign capital became less significant relative to domestic capital, all these regimes faced pressures for democratization (and have become more democratic) in order to assure business, labor, and private savers that their interests would be taken into account, and to avoid debilitating capital flight.

China's regime is far from unified, and has a history of hostility to, not cooperation with, private entrepreneurs. In addition, the regime cannot hope for export-oriented foreign investment to provide for more than a fraction of China's development needs. For example, Taiwan gains approximately $3,000 per capita per year from exports, providing almost one-half of GNP. For China to gain even $500 per capita per year through exports in 2010 would require China's total exports to reach $700 billion, almost three times Japan's total exports today. Even this modest per capita gain may be out of reach. While foreign investment in export-oriented manufacture will transform a sliver of China's population living along the coast, domestic capital from private savings will have to be tapped to reach the bulk of China's population.[15] Even the examples of Korea, Taiwan, Brazil, and Mexico argue that as China grows more dependent on domestic private capital and investment, business leaders and workers will demand increased democratization and responsiveness from the regime. Indeed, given the divisions in China's leadership, the traditional hostility of the Party to private business, and the lack of natural resources or foreign capital available to the regime, the ability of the Party to orchestrate an authoritarian development drive is highly doubtful.

Given that greater market freedom offers the only escape from China's economic difficulties, and that China's leadership divisions and elite conflicts mean that such greater market freedom will inevitably produce demands for greater political freedom and democratization, the only question is whether the transition to greater market freedom and democratization can occur relatively peacefully, as in Czechoslovakia, Hungary, and East Germany, or whether it will involve internal war and ongoing political crises, as in the U.S.S.R. and Yugoslavia. To pose the question in these terms is almost to provide an answer. Czechoslovakia, Hungary, and East Germany are relatively small nations, with prior democratic experience and/or strong cultural and political ties to firmly democratic western Europe. Communist China, like the U.S.S.R. and Yugoslavia, is a multiethnic, multireligious empire, including Moslem lands to the west and Tibet to the south, which have a history of rebellion against the central authority in Beijing. In addition, China, like the U.S.S.R., has no democratic history of its own, and has even scantier ties to the West. Other than traditional loyalties—to home regions, to family, and to the desire for material improvement—there is little to hold together the Chinese state if the Communist ideology and political superstructure should fail. Much as in the U.S.S.R., the likely result of a confrontation between forces supporting private enterprise and conservatives supporting the Communist Party is a dissolution of the present Chinese state into a number of independent regional entities.

Both the history of China, and the more general history of regime crises, indicates that regime collapse is highly likely to lead to international war. Both revolutionary and counterrevolutionary factions often seek international intervention to support their position, either inviting foreign powers to send in their forces or pursuing campaigns against perceived enemies abroad.[16] In East Asia, the conflict between North and South Korea may be sharply affected by political disorders in China; the loss of Chinese influence may also lead to intensified problems in Cambodia. Just as the collapse of the U.S.S.R. unleashed conflicts in Yugoslavia, so the collapse of Communist China may unleash conflicts in Southeast Asia or the Korean peninsula.

Moreover, as with the U.S.S.R., the likely fissioning of post-Communist China into a number of new states will create new problems. At the least Tibet is likely to seek international assistance for its independence. In addition, Sichuan—traditionally a breadbasket province, and one with a long history of separatist rebellion—may, like the Ukraine in Russia, seek independence, imposing great food strains on other regions. Conflicts for control of the Yellow, Yangzi, and other major rivers, and for control of weapons and resources, may add to the existing tangle of regional and ethnic conflicts that already divide China's provinces. Once the glue of unified Communist rule dissolves, China may once again, as it has so often in its history following the fall of unifying dynasties—experience a decade or even century-long interregnum of warring among regional states.

There is a possible alternative path of events. We have noted that the challenges facing the regime from declining resource/population balances are inescapable. Adjustment to changing ecological conditions will mean large-scale migration, rapid urbanization, and increases in private industry. Even with these changes, given China's vast population, its low income starting point, and its low ratio of resources to population, the living standards of most Chinese are liable to continue to lag far behind most of their Asian and Southeast Asian neighbors, while regional, urban/rural and cross-class income inequalities are likely to increase. Unlike the easy economic gains that came from Deng's abandoning the worst aspects of Mao's utopian socialism, future gains will likely be more difficult, more uneven, and create greater inflationary pressures and employment dislocations. To meet these challenges and survive, China's leadership will require increased popular support. Contrary to Deng, the Party may benefit if it seeks such support through democratizing reforms.

If there is a lesson to be gained from the comparative experience of the U.S.S.R., Hungary, and the post-Soviet states, it is that Communist leaders gained more stable regime transitions when they chose to seek democratic support before economic conditions deteriorated so far that the Communist leadership could not escape blame. That is, in Hungary, Ukraine, and several Central Asian post-Soviet states, former Party officials have retained a share, sometimes the predominant share, of power by accepting the need to relinquish the Party's effort at absolute control, and instead to seek renewed legitimacy by competing in democratic elections. It remains to be seen whether such democratization proves an unstable sham, or a solid foundation for stable democracy. However, it seems that the path China is determined to follow—largely abandoning Communist management of the economy while maintaining tight Communist Party control of society—is destabilizing.

The key example is the U.S.S.R., where Mikhail Gorbachev sought and failed to retain Party control while struggling with economic reform. In the U.S.S.R. too, changing ecological conditions and economic stagnation—in this case, environmental degradation and economic failures so severe as to depress life expectancy and dramatically raise infant mortality—produced widespread pressures for reform, both within and outside the Party.[17] Gorbachev attempted to reform the Party and use it to implement economic changes, while maintaining the Party's monopoly of power. This strategy allowed Party conservatives to sabotage change from within, while Gorbachev and the Party could not escape blame for the continued deterioration of the Russian economy. Attacked by conservatives for seeking change, and attacked by reformers for being unable to produce it, the authority of both Gorbachev and the Party crumbled. If China's aging Communist leaders attempt to follow Gorbachev in seeking to limit the devolution of Party power, they may follow him into oblivion as China's economy creaks under the strain of accommodating its growing population.

On the other hand, if China's Communist Party, having already all but abandoned Communism, were to seek to relegitimate itself through moving to implement democracy, it might provide more long-run stability for its role. There will be costs; Tibet may seek autonomy, as may the Muslim lands of central Asia. The Party will inevitably split into conservative and reform wings, which will contest elections. The army will have to be neutralized as a political arbiter. Nonetheless, the example of several Eastern European countries suggests that if reformers in the Party abandon efforts at maintaining direct Party control, form an alliance with business leaders, and seek democratic support, they are more likely to gain a stable transition to a market economy than if they attempt to free the economy while maintaining Party control and reforming the Party from within. The latter course, as Gorbachev discovered, allows conflicts within the Party to paralyze change, while leaving the Party as a whole to take all the blame for continued economic and political stalemates.

In sum, long-term stability is more likely to be achieved by taking the plunge into democratization early, while Party leaders have some credibility and support for the positive elements of Deng's reform program, and before economic setbacks, struggles over internal migration, and problems of regional conflicts undermine the Party's remaining support. A planned, gradual phase-in of democratic institutions may create short-term uncertainties. However, aiming for short-term stability by maintaining Party control is more likely to lead to the kind of long-term collapse and chaos now evident in the former U.S.S.R. and Yugoslavia, and all too easy to envisage in China.

Conclusion

China's environmental problems will force major economic adjustments and reforms. Since such reforms are likely to intensify the mounting tempo of confrontations among Party factions, elites, workers, and other groups that have erupted every few years for the last decade, and given that the Communist leadership appears unwilling to grant the democratic reforms that might gain it renewed support, we can expect a terminal crisis within the next ten to fifteen years.

In Europe, policy planners have learned that the end of the Cold War did not mean an end to international security threats; it merely meant the rise of new, unforeseen, and largely unanticipated security threats from the chaotic collapse of former Communist regimes. Even while welcoming the likely demise of Communism in China, it is necessary to begin envisioning responses to the international security issues that are sure to arise in the wake of China's transformation.

--- **NOTES** ---

1. The population data in this paragraph, and the appraisal of current trends in arable and environmental resources, are based largely on Vaclav Smil, *China's Environment* (Armonk, N.Y.: M. E. Sharpe, 1992).

2. P. Liu and K. Hwang, "Population Change and Economic Development in Mainland China since 1400," *Modern Chinese Economic History*, C. Hou and T. Yu, ed. (Taipei: Institute of Economics, Academica Sinica, 1977), pp. 61–81; K. Chao, *Man and Land in Chinese History: An Economic Analysis* (Stanford: Stanford University Press: 1986), pp. 85–89.

3. J. Shang, "The Process of Economic Recovery, Stabilization, and Its Accomplishments in the Early Ch'ing," *Chinese Studies in History* vol. 15 (1981–1982), p. 25.

4. D. D. Perkins, *Agricultural Development in China, 1368–1968* (Chicago: Aldine Press, 1969), p. 16; G. Wang, "The Chinese Urge to Civilize: Reflections on Change," *Journal of Asian History* vol. 18 (1984), p. 7.

5. P. Perdue, *Exhausting the Earth: State and Peasant in Hunan, 1500–1850* (Cambridge, Mass.: Council on East Asian Studies, 1987), p. 88.

6. S. M. Jones and P. A. Kuhn. "Dynastic Decline and Roots of Rebellion," *The Cambridge History of China* vol. 10: *Late Ch'ing, 1800–1911, Part 1*, D. Twitchett and J. K. Fairbank, eds. (Cambridge, U.K.: Cambridge University Press, 1978), p. 109.

7. G. W. Skinner, "Sichuan's Population in the Nineteenth Century: Lessons from Disaggregated Data," *Late Imperial China* vol. 8 (1987), pp. 1–79; Wang, op. cit., p. 7.

8. The broader patterns of population pressure and political conflict throughout Eurasia from 1500 to 1850 are covered in greater depth in J. Goldstone, *Revolution and Rebellion in the Early Modern World* (Berkeley and Los Angeles: University of California Press, 1991).

9. B. Ashton et al., "Famine in China, 1958–61," *Population and Development Review* vol. 10 (1984), pp. 613–45.

10. A. G. Walder, "The Political Sociology of the Beijing Upheaval of 1989," *Problems of Communism* (September/October 1989), p. 31.

11. T. McDaniel, *Autocracy: Modernization and Revolution in Russia and Iran* (Princeton, N.J.: Princeton University Press, 1991).

12. J. T. Dreyer, "The People's Liberation Army and the Power Struggle of 1989," *Problems of Communism* (September/October 1989), pp. 41–48.

13. A. G. Walder, "The Political Sociology of the Beijing Upheaval of 1989," op. cit., p. 35.

14. A. G. Walder , "The Political Sociology of the Beijing Upheaval of 1989," op. cit., p. 37.

15. The per capita GDP and import figures for China, Taiwan, and Japan are taken from *World Almanac* (New York: Pharos, 1992), pp. 748, 773, 806. China's capital outflows are estimated by Richard Hornik, "Bursting China's Bubble," *Foreign Affairs* vol. 73 (1994), pp. 28–42.

16. S. M. Walt, "Revolution and War," *World Politics* vol. 44 (1992), pp. 321–68.

17. M. Feshbach and A. Friendly, Jr., *Ecocide in the U.S.S.R* (New York: Basic Books, 1992).

11

Out of Focus

U.S. Military Satellites and Environmental Rescue

Ronald J. Deibert

From the first crude accounting records of ancient Sumeria, information technologies have provided human beings with the means to monitor and record their relations with each other and their environment. Control over these technologies has proven to be an important source of social power. This capacity has varied considerably over time with changes in the form of communications. Today, digital computers, fibre optics, and earth-orbiting satellites have provided the tools to monitor human relations on a planetary scale, a capacity that is vital for environmental rescue. The subject of this chapter is the control and use of one important component of this planetary information web—satellite reconnaissance.

Space-based monitoring of the earth plays a central role in all facets of environmental research. The obvious advantage of using satellite reconnaissance is the ability it provides to monitor the whole planet in a variety of spectral modes and resolution capabilities. Given the scope of problems to be addressed, space-based surveillance will likely be an integral technological component of global environmental governance well into the future. As one environmental researcher put it: "Electronic and optical technologies of every kind will underpin what will ultimately be a vast orbiting and terrestrial infrastructure for monitoring and modeling the global climate and environment."[1] Control over these systems will consequently have a strong bearing on the nature and direction of environmental rescue.

Although a multinational community of civilian environmental researchers gradually emerged over the last few decades to exploit satellite reconnaissance, it was military/intelligence organizations that initially provided the capital and the impetus to develop this technology. Spurred on by their mutual hostility, the United States and the Soviet Union invested billions of dollars and countless hours of research into the creation of sophisticated space-based reconnaissance systems designed to monitor and target adversary force structures. For many years, while most of the world's familiarity with satellite reconnaissance was limited to the meteorological images displayed on nightly weather forecasts, the two superpowers maintained a monopoly on a vastly superior technology. Little was revealed about the capabilities of such systems and virtually nothing was shared. For years, satellite reconnaissance was obliquely referred to as "national technical means."

After the collapse of the Cold War, the military found it difficult to sustain this shroud of secrecy. Pressure has been placed on the intelligence agencies responsible for satellite reconnaissance to contribute to other areas, including the environment. The logic of such restructuring seems to make intuitive sense: not only is the military a source of untapped expertise, but military satellite reconnaissance systems are among the most advanced technologies available. Technically speaking, the only real distinction between civilian and military satellites is in terms of the level of image sophistication, so refocusing those satellites to environmental ends would require only a change in mission. Environmentalists gain needed resources and expertise while the military finds a new rationale for continued funding. A natural compatibility is born!

Yet the issue of refocusing military satellites to environmental ends touches at the heart of what has become known as the "environmental security" debate. Should the military be given a role in environmental rescue? What are the risks of redirecting military expertise toward the environment? Are environmental problems and military solutions compatible? In short, the issues raised above point to normative concerns regarding the appropriate channels in which energy and expertise should be directed in the service of environmental rescue.

After briefly reviewing the contending perspectives on the relationship between the "environment" and "security," this chapter focuses on the question of redirecting U.S. military satellites toward environmental protection in the context of the recently created Strategic Environmental Research and Development Program (SERDP) and other similar initiatives. The institutions that have evolved for both military and civilian approaches to satellite reconnaissance are outlined and compared. The argument made here is that the case of U.S. satellite reconnaissance offers a clear illustration of the perils of redirecting military expertise toward the environment. This argument rests on the belief that military and civilian approaches are incompatible in fundamental ways. Rather than turn to the military, environmentalists should instead promote environmentally dedicated systems and further international cooperation.

Environmental/Security Issues

As environmental problems mount, the linking of the "environment" with "security" has been more common as observers search for a way to characterize aptly the weighty issues involved. Although environmental security is now widely recognized as a subdisciplinary focus among international relations theorists, the linking of the two terms has spawned a contentious debate. While it is beyond the scope of this chapter to provide a lengthy discussion of the details of this debate, three contending positions will be outlined. Each of these positions offer different interpretations of the relationship between environmental degradation and security. More importantly for this chapter, each interpretation, in turn, has important institutional ramifications for global environmental governance.

The first group is made up of those who contend that the environment has little or no impact on questions of national security, either because the dynamics of international security are thought to have an autonomous logic apart from the supposed "low" politics of the environment or economics, or simply because environmental problems are assessed as being relatively insignificant for military/strategic issues.[2] This group of security purists is largely made up of military analysts and strategic studies experts, and probably represents a majority within the *security* studies field, yet an increasing minority outside. For this group, *security* should not be redefined to incorporate nontraditional, environmental problems, nor should military organizations be redirected to environmental ends.

A second group is made up of those who, on the contrary, believe that environmental problems represent a significant new "threat" to national security.[3] Theorists in this second group buttress their claim by pointing to studies that identify the potential for violence as resources become scarce in the developing world.[4] Consequently, these observers advocate redefining *security* beyond its traditional military scope to include environmental-related issues. From this perspective, "security threats" are adjusted while the focus of protection—the nation-state—remains the same. In a quote representative of this position, Peter Gleick outlines the way in which environmental conditions will structure national security responses:

> Because of these problems a nation or region bent on protecting its "security" in the future will have to concern itself as much with the flows of the planet's geophysical capital as it does today with the flows of economic capital; as much with the balance of atmospheric trace gases as with the balance of military power; as much with monitoring the earth's vital signs as with monitoring the arsenals of destruction.[5]

The institutional ramifications of this second position are clearly inclined toward redirecting military organizations and expertise to "fight" environmental "threats." Such a redirection is thought to have a double benefit: first, by emphasizing the severity of environmental degradation, particularly its latent potential for military conflict, environmental issues are thus into the higher echelons of national decisionmaking priorities—into the realm of "high" politics. Environmental issues thus receive first-order consideration. Linking the environment with "security" in this way can be seen as a discursive strategy to tap into the energies and fears normally reserved for war. Consequently, a common feature "is the use of language traditionally associated with violence and war to understand environmental problems and to motivate action."[6] Second, by redirecting military organizations and expertise to environmental ends, a more humane purpose is thought to be served. Military institutions are focused on protection, rather than destruction, of the environment. The power of this appeal is amplified today when military organizations and the private corporations that profit from them face considerable budgetary pressures and restructuring tasks.[7] In the face of declining military threats, adaptation to a new role, rather than outright conversion, is clearly a more attractive option for vested interests of the military-industrial complex. The SERDP (which will be outlined in more detail below) is a clear example of the type of project championed by this second group.

A third group of theorists, while sharing the belief that environmental issues are a serious concern, reject many of the assumptions and metaphors of the second group—in particular, that environmental problems can be tackled within the existing states system framework. According to this group, a wholesale revision of the current international order is essential if human beings hope to adapt successfully to ecological change. Consequently, the very institutions that the second group wishes to reorient, the third group wishes to overturn. In Deudney's opinion, "instead of linking national security to the environment, environmentalists should emphasize that global ecological problems call into question the nation-state and its privileged status in world politics."[8] At the heart of this argument is the recognition that environmental problems do not fit neatly into the segmented territorial boundaries of the modern states system; consequently, if planetary habitability is to be ensured, state autonomy will have to be sacrificed to a higher level of governance. As Dalby points out:

> Pollution knows no boundaries. As a result, the communities to which appeals for ecological security must be made are ones that are not clearly geographically bounded. Ecological regions do not necessarily coincide with territorial boundaries; questions of sovereignty are inevitably raised by ecological issues.[9]

There are divisions within this last group over the appropriate conceptualization of this transformation. While there is agreement that the states system and planetary governance are incompatible, many moderate theorists within this group argue that the way to conceptualize the new order is through a broadening of the notion of security outward to include the entire planet in the form of "common" or "comprehensive" security.[10] Others are critical of appropriating the term *security* at all, arguing that by doing so environmentalists also risk appropriating the mindsets and institutions associated with the term. For example, security is often identified with maintaining the status quo—something that is clearly incompatible with the core mission of this group: transforming the states system.[11] For Deudney, the term *security* has too strong a link with nationalism and war. Rather than adopt military language, environmentalists should "directly challenge the tribal power of nationalism and the chronic militarization of public discourse."[12]

At issue is more than just semantics. As Conca explains, security metaphors can "shape and limit how we conceive of problems and solutions" and by doing so, might also "privilege the role of military institutions in policy responses."[13] From the perspective of the third group, the contradictions and risks of any military involvement are clear: modern military organizations are an integral component of the states system. They are primarily charged with defending the state and preserving territorial integrity—the very same institutions that the third group is working to overturn. Incorporating the military and the language of war in environmental rescue only perpetuates organizations and mindsets that must be overcome if environmental rescue is to be successful. Though security metaphors are a matter of contention, for the third group as a whole military institutions should not be given a role in environmental response.

The debate outlined above is made all the more weighty given that issues of environmental degradation are being thrust into the spotlight at precisely the same time as military and intelligence organizations are facing cutbacks and restructuring assignments. Hearings have been held in both the House and Senate Select Committees on Intelligence to oversee the restructuring of the intelligence community.[14] New roles have been considered for intelligence missions, including the seemingly contradictory assignments of environmental protection and national economic intelligence (said by one important observer to include ensuring access to vital raw materials).[15] How these conflicting objectives (state security, environmental rescue, and economic support) will be reconciled in practice is unclear. Nevertheless, considerable initiatives are under way to enlist defense and intelligence expertise in support of environmental research. The remainder of this chapter will explore the details of one such program by focusing on the case of U.S. military satellite reconnaissance and the environment in the context of the SERDP.

Satellite Reconnaissance and Environmental Security

Earth observation is vital to the operations of military/intelligence and civilian/environmental organizations, though for completely different reasons. While the tools are similar, the purposes for which they are deployed are not. As a consequence, the institutions and work ethics that have evolved for each reflect these differing underlying interests. At issue in this chapter is whether or not U.S. military/intelligence institutions and work ethics can be reoriented to environmental ends without at the same time undermining those very ends. To set the context, I now turn to a more detailed investigation of both U.S. military/intelligence and global civilian/environmental satellite reconnaissance operations.

U.S. Military Satellite Reconnaissance Systems

Reconnaissance has always been an important ingredient of military operations, though the capabilities have varied considerably as technology has evolved. It was not until after World War II with developments in ballistic missile technology that serious consideration could be given to the idea of a space-based reconnaissance platform. Studies were undertaken by the RAND corporation in the 1940s and the 1950s envisioning a number of alternative designs, but it wasn't until 1960 that the first satellite reconnaissance system became operational.[16] In the same year, the National Reconnaissance Office (NRO) was created to coordinate the satellite programs of the Air Force, Navy, and Central Intelligence Agency (CIA). Since then, the NRO has been the central organization involved in U.S. military satellite reconnaissance.

The formidable U.S. space armada consists of a variety of different Earth-observation systems, each with unique capabilities designed for specific military-intelligence tasks. The optical satellite program is normally referred to as "Keyhole," or KH for short, and includes a series of satellites designed primarily for high-resolution imaging. What exactly can be "seen" by such satellites is the subject of considerable public speculation. While exact specifications are classified, it is widely known that the most advanced KH-11s can resolve objects as small as a dozen centimetres.[17] The optical satellites are complemented by: the LACROSSE series of synthetic aperture radar (SAR) satellites, which provide images in all-weather/day-night conditions; sophisticated electronic eavesdropping satellites that monitor various forms of communications, from the interception of radar signals to computer and telephone traffic; early-warning satellites designed to detect the electromagnetic pulse of a ballistic missile launch; and the NAVSTAR Global Positioning System (GPS), which is used for navigation purposes, providing soldiers with precision details of ground location within fifteen metres. Binding all of these systems together is the Defense Communications System (DSCS)—a series of high-powered communications satellites that relay the collected data to ground-based nodes.[18]

This complex of space-based information systems forms the vital backbone of U.S. military operations today. No better indication can be found of the U.S. reliance on satellite reconnaissance than the Gulf War, where the United States operated as many as eight spacecraft continuously throughout the conflict.[19] The satellites were used to record targets, coordinate attacks, and assess damage prior to and during the conflict, imposing a constant transparent surveillance grid over Iraq. Another indication of their importance is the massive annual budget of the NRO, traditionally the second largest in the entire intelligence community behind only the National Security Agency. In the past, the United States has budgeted as much as $6 billion annually on the NRO's operations, with each satellite costing as much as $1 billion to build.[20] It is estimated that as much as $150 billion in total has been spent on military space operations over the past forty years.[21]

The missions and targets of U.S. satellite reconnaissance are essentially derived from a perception of threats at a particular time. Given that the motivation to develop satellite reconnaissance grew directly out of a need to monitor military forces and intentions, it is not surprising that for most of their history these systems were primarily focused on the Soviet military.[22] Periodically, however, they also were used to monitor other nonmilitary phenomena, including environmental conditions. For example, during the Cold War satellites regularly monitored Soviet crop production. They have also been used to monitor migration and refugees stemming from ethnic conflict, and more recently have played a part in fighting the "war on drugs."[23] Mexico officially protested the U.S. use of satellite reconnaissance to monitor narcotics crop production after the State Department issued its annual report on foreign narcotics estimating that 143,133 acres were being cultivated for marijuana—a figure one author describes as estimated with "eerie precision."[24]

Evolutionary changes in the international system, coupled with the abrupt end of the Cold War, have placed considerable pressures on the NRO to adapt to new conditions. Certainly, national, regional, and commercial proliferation of satellite reconnaissance systems around the globe challenge the NRO's predominance in space-based surveillance operations.[25] As sophisticated imagery becomes more widely available, the NRO loses considerable leverage—a threat that has not gone unnoticed by top-ranking officials, as will be explained below. Probably the most immediate pressures, however, have been budget contractions. From an annual high of around $6 billion, the budget for the NRO dropped to $3 billion in 1991 and to $2.5 billion in 1992. The NRO has responded to these pressures by searching for new targets and threats. Despite the collapse of the Soviet Union, officials have publicly complained of "high demand" and an "explosion of requests for a broad range of intelligence data."[26] Although it is likely that the priorities of U.S. satellite reconnaissance will remain of a military-strategic nature for some time to come, the temptation to engage in other nonmilitary monitoring tasks will be strong as budgetary and public pressures converge on the NRO's privileged operations. The SERDP, and other initiatives like it, may be but an early precedent of the type of crossover nonmilitary monitoring mission engaged in by the NRO to justify continued funding.

Civilian Satellite Monitoring of the Environment

Applications of satellite reconnaissance are not restricted to the military/strategic missions outlined above. Indeed, environmental satellites have been used in Earth-observation activities for many years and are developing rapidly today. Just as the military threats of the early post–World War II period spurred development of military/intelligence reconnaissance, the widespread recognition of environmental degradation has generated multinational interest in the development of environmentally dedicated satellite systems. Additionally, the collapse of the Cold War has removed a considerable political barrier to the dissemination of sophisticated technologies outside of strict national security circles.

Civilian environmental satellites currently operate under both national and international direction, and are coordinated by several informal international regimes. The most important of the informal international bodies are the Committee on Earth Observation Satellites (CEOS) and the Earth Observation International Coordination Working Group (EO-ICWG). CEOS and ICWG work alongside regional bodies, like the European Space Agency (ESA) or the European Organization for the Exploitation of Meteorological Satellites (EUMETSAT), as well as national organizations, like the National Aeronautics and Space Administration (NASA) and the National Oceanographic and Atmospheric Administration (NOAA) in the United States.[27]

As global environmental problems have mounted over the last two decades, there has been strong motivation to intensify international cooperation in Earth observation and to move away from purely autonomous national programs. This represents a considerable change from initial program developments at the beginning of the space age, when aggressive national competition under the shadow of the Cold War fueled research and development. Though competitive commercial pressures still often act as a barrier, significant movement has been made in coordinating existing and future national and regional programs to avoid overlap and to ensure international distribution of data among scientists. The 1992 International Space Year (ISY) is a significant landmark in this respect, drawing together twenty-nine countries and ten international bodies in a wide variety of programs under the theme "Mission to Planet Earth."[28] There has also been a spate of policy proposals envisioning future large-scale, globally organized Earth-observation activities. While it is unlikely that most of these proposals will move beyond the conceptual stage, the sheer number of them made reflects a growing interest in international coordination of civilian environmental satellites.

By far the most ambitious of these proposals is the planned Earth Observing System (EOS), a multiyear $8 billion NASA-directed initiative to study the earth's biosphere in all its aspects.[29] The centrepiece of the U.S. Global Change Research Program, EOS is considered by researchers to be the prototype of future long-term monitoring missions, and like most other NASA programs the EOS will entail significant multinational participation and coordination. The current status of the EOS has been significantly trimmed from the original, somewhat grandiose, early plans. Initially, the EOS was to involve two 30,000-pound massive space platforms, with new platforms being launched every five years costing as much as $30 billion.[30] After these initial plans came under criticism, a modified program was put in place involving two intermediate platforms plus a series of small satellites to be launched in the 1998–2010 time frame, coordinated by a terrestrial infrastructure called the Earth Observing System Data and Information System (Eosdis).[31] Other similar

Earth-observation platforms are being developed by Japan and Europe, and are coordinated through the ICWG.[32]

Once operational, these proposed systems will complement a wide variety of environmental satellites currently in orbit. Meteorological satellites, like the GOES and POES series of weather satellites and a number of NASA and NOAA-directed atmospheric systems, provide information on everything from climate patterns to ozone depletion. Nonmeteorological commercial satellite systems include the U.S. LANDSAT and the French SPOT satellite series, and the European ERS-1, the Japanese JERS-1, and the Canadian RADARSAT SAR satellites. Together, these systems perform a wide range of environmental missions, including monitoring wind height and wave speed over oceans, aiding the management of territorial ecosystems through surveillance of vegetation and rainfall patterns, pinpointing fertile farmlands for refugee resettlement, monitoring and detecting ecological and natural disasters, and urban planning.[33] Though satellite reconnaissance was first developed by the military, environmentally dedicated systems are proliferating rapidly in response to the environmental crisis.

Declassifying U.S. Imagery for Environmental Support

The idea to exploit military systems for research seems intuitively attractive. Indeed, *technically* speaking, those hoping to channel defense expertise into environmental missions would find satellite reconnaissance to be one of the most amenable technologies since most of the current civilian applications were an outgrowth of military innovations. The same type of reconnaissance platforms that were developed by the military to distinguish between camouflage and growing vegetation are also used by resource managers to make estimates of crop yield. It is only natural, then, that pressures for military reorientation should bear strongly on the NRO.

Congressmen George Brown and Sam Nunn, and former Congressman (now Vice President) Al Gore spearheaded the formal public campaign to enable environmentalists to have access to spy satellite imagery and also to redirect efforts at the NRO toward environmental ends. The reorientation of satellite systems is subsumed under the general SERDP, a program specifically devised to exploit military technology and expertise for environmental missions. Invoking the Strategic Defense Initiative as a model, Gore initially proposed in 1990 the launch of the Strategic Environment Initiative, arguing that "the environment has become a question of national security" requiring the "mobilization of talent and resources on a scale ordinarily reserved for national defense."[34] Speaking before the Senate in June 1990, Nunn gave three reasons for his proposed SERDP:

First, because environmental deterioration in a very real sense threatens our nation's security and the security of the world; second, because the defense establishment has unique data collection and technological capabilities; and third, because the defense establishment helped create some of the environmental problems we face today.[35]

Although $200 million in funding was proposed by Nunn for SERDP, the actual budgetary figures have been considerably less: $77 million was appropriated for FY1991; $70 million for FY1992; $180 million for FY1993; and $164 million for FY 1994. Projects relating to space-based remote sensing have been included in the SERDP over the last four years, the largest of which is the Global Change and Remote Sensing project jointly managed by the Department of Energy and the Department of Defense. The project contains provisions to exploit classified and recently declassified defense expertise and technology in remote sensing for studies of ozone depletion and global warming.[36] Two assignments directly related to satellite reconnaissance include an $850,000 project to create an archive of meteorological data from the Air Force's Defense Meteorological Support Program (DMSP) and the release of ocean surface satellite data from the Navy's Geodesy satellite.[37]

Other non-SERDP programs to exploit classified data exist, though little can be revealed about them. The Council on Foreign Relations sponsored a private discussion between the NRO and a select group of environmental scientists in 1992 to help devise appropriate procedures for declassification of data—an extremely sensitive issue for intelligence officials.[38] Of course, scientists cannot roam freely through NRO archives. Most likely, closed-door meetings of this sort will be the only type amenable to intelligence officials when distribution of classified data is involved. To date, the procedure has been that scientists make requests for data "blindly" with intelligence officials determining whether or not to grant access. More recently, the NRO has taken on a more active reorientation toward environmental missions, with spy satellites actually being deployed to monitor specific environmental phenomena around the world.[39] However, the data derived from these missions remains classified and will only be released to "future generations" of environmental scientists.

From a purely technical perspective, NRO satellite imagery can provide a unique contribution to environmental monitoring. The far better resolution of these satellites could provide information not available from existing commercial imagery. As Richelson points out, "a measure of the utility of the high-resolution imagery lies in the fact that CIA estimates of Soviet crop yields have been consistently more accurate than those of the Department of Agriculture, The much greater speed by which these systems operate could be used during disasters to pinpoint fires or floods in near real-time, and through superimposition on a map give instantaneous data."[40] An even greater benefit is foreseen through access to the vast archived images. Commercially available images became available only in the mid 1970s, and did not develop fully until the 1980s. Archived images from U.S. reconnaissance systems date back to the early 1960s. The same images that were taken of a particular missile site in the Soviet Union over a thirty-year period might also help to chart snowfall in that particular region. Gore has recalled how archived images were helpful in pinpointing old hazardous dump sites in west Tennessee.[41]

Other analysts are less sanguine about the benefits to be gained from NRO satellite imagery. An obvious limitation of the keyhole satellites is their narrow field of view. Unlike civilian satellites, which have a wide swath width, optical spy satellites have finer resolution limiting the geographic area of the image, a feature that has been likened to seeing the world through a soda straw.[42] Others have noted that since the data was not collected for scientific purposes, it may lack the necessary consistency and reliability.[43] The data that has already been released has received mixed reviews from the scientific community. A good portion of the data may be completely redundant considering that environmentally dedicated systems are already well developed in certain areas. One example is the ocean surface data released by the Navy's satellites, which are virtually identical to that produced by the ERS-1 and RADARSAT SAR satellites.[44] While current debate has centered on the purely instrumental utility of the images as outlined above, little attention has focused on the larger ramifications of redirecting intelligence agencies toward the environmental arena. As shown below, there are significant hazards in such an effort that should merit considerable scrutiny by environmentalists.

From Deep Black to Green?

The positions of those who are opposed to redirecting military expertise and technology to environmental ends were outlined earlier in the chapter. At the heart of such criticism is the belief that groups committed to global habitability and groups committed to national security suffer conceptual and organizational mismatch. As Deudney notes:

> Organisations that provide protection from violence differ greatly from those in environmental protection. . . . Military organisations are secretive, extremely hierarchical and centralized, and normally deploy vastly expensive, highly specialized and advanced technologies.[45]

Perhaps no better illustration of Deudney's concerns can be found than the organizations surrounding U.S. military reconnaissance. The NRO is, of course, one part of an institutional complex that together comprises the entire U.S. intelligence community. The various organizational components of this community share certain deeply engrained operational norms and ethics that have evolved over the course of the Cold War, and have become routinized into habitual daily practice. Perhaps the most pervasive norm among the intelligence community is the widespread secrecy that informs all operations—and in this respect, the NRO is paramount. Satellite reconnaissance is governed by a system of classification known as Sensitive Compartmented Information, or SCI—a classification that Burrows describes as being even "blacker" than the top-secret category.[46] The NRO is so deeply imbued with classification that its very name and mere existence was not officially revealed until 1992. Those who are employed by the NRO maintain a close-knit inner sanctum—a community in the intelligence shadows. Burrows explains that "there is a kind of reconnaissance club, an unofficial secret society composed of 'black hats' from the various contractors, military services, and the intelligence agencies and divisions, all of whom carry the appropriate clearances and are scrupulous about remaining in deep shadow."[47] Levels upon levels of classification schemes compartmentalize information within the NRO itself, and between the NRO and the "outside."

The pervasiveness of secrecy throughout the NRO has smothered more than one past attempt at declassification. Following the initiation of the Strategic Arms Limitation Talks in the 1960s, mention was made of using satellites for verification. The NRO adamantly opposed the idea, arguing that using such systems for arms control purposes would require the United States to make its capabilities public to demonstrate that it could verify compliance.[48] A good indication of what environmental researchers might expect can be seen in the experiences of the U.S. Geological Survey (USGS). During construction of the Alaskan pipeline, initial arrangements were made between the USGS and the NRO to allow workers to have access to images that would give them a detailed overview of the terrain. After sifting through the NRO's declassification schemes, the images that were finally released were so devoid of any standard topographical markings found on satellite images that they were virtually useless to the USGS.[49]

This extensive institutionalized secrecy could conceivably present a problem if, for example, significant overlap exists between imagery intended for environmental purposes and imagery intended for national security and military strategy. A proper analysis of data might be hampered by such secrecy, especially if gaps in data result from overzealous NRO classification. Even more troubling would be the possibility of deception or manipulation of data to serve national security interests that might not coincide with environmental research and protection. The history of U.S. intelligence as a whole is marked by incidents of public duplicity and subterfuge to such an extent that manipulation of environmental data would not be out of character. Describing the operational objectives of the NRO, former Director Martin Faga has said that "the perfect world is where we have all the information and he, whoever the adversary is in this context, has none and knows nothing about what we know."[50] Such an ethos, though justified from a military planner's perspective, is fundamentally incompatible with environmental scientific research, which is premised on open exchange of information.

Defense Conversion or Cooptation?

A second, related concern is the encroachment of the military into existing civilian satellite systems. While attention has naturally focused on defense conversion in the aftermath of the Cold War, little corresponding attention has been given to what might be called military cooptation; that is, the use of environmental satellites by the military for strategic purposes. Should environmentalists rely more on the military for data, the possibility exists that there will gradually be no alternative to military monitoring—that the military will become a "clearing house" for environmental data. The perils of such concerns are clearly illustrated by the policies of the French SPOT civilian satellite during the Gulf War. Generally, environmentalists have benefited from SPOT's "open-skies" policy, allowing the purchase of images virtually anywhere on Earth. That policy came to an abrupt end, however, with the onset of the Persian Gulf War. After the Iraqi invasion of Kuwait, SPOT-image imposed a blackout of images of the Persian Gulf. One notable exception was the U.S. Pentagon.

During the Gulf War, the U.S. Air Force became the single largest consumer in the world of commercial satellite imagery. Not only did the images provide wide-area coverage not available from the KH-11 images, but data derived from the images were used to program "smart bombs" and to orient flight routes and bombing missions through simulations prior to assaults.[51] Images from civilian satellites like SPOT and LANDSAT are regularly fused in this way with more sophisticated satellite imagery to train students and to plan invasions. For example, the U.S. Department of Defense's TENCAP (tactical exploitation of national capabilities) simulation system has been used to plan the invasion of Panama and the relief mission over Bosnia-Herzegovina.[52] As one Air Force General noted ominously about the military's recent exploitation of civilian systems: "We have imagery for the entire world."[53]

The military/strategic uses of environmental satellites have already affected the operational priorities of such systems. Though SPOT-image insists that most of its clients are from the resource management and civilian sectors, much of its advertising campaign is geared to exploiting the military market. Its advertisements offer "a new way to win."[54] In a promotional booklet entitled "Surveillance," SPOT-image described the way in which their images could assist military strikes:

> Before taking any decision concerning a target located deep inside a zone inaccessible to reconnaissance planes . . . three main points need to be analyzed: the target, the "threats" surrounding it, and the route leading to it. The ideal way to handle these studies is to observe SPOT satellite images.[55]

Further cooptation can be seen in trends in the production of new satellite systems. Increasingly, economic efficiency arguments are being used to push for the consolidation of military and civilian/

environmental research, development, and operations. As one U.S. Office of Technology Assessment report argued: "This increased use of CTIB [commercial technology and industrial base], dubbed civil-military integration (CMI), can take many forms, including purchasing commercially available goods and services, conducting both defense and commercial research in the same facility, manufacturing defense and commercial items on the same production line, and maintaining such items in shared facilities."[56] For example, until an interagency squabble between NASA and the U.S. Department of Defense over system design derailed coproduction, Landsat 7 was to have been operated as a joint civilian-military satellite.[57] Likewise, in May 1994, the Clinton administration announced a proposal to "consolidate" the polar-orbiting meteorological programs of NOAA and the Department of Defense and to coordinate both to avoid overlap with the programs of NASA's EOS.[58] More recently, several U.S. corporations won a prized contract to build the Brazilian Amazon Surveillance System (what Bill Clinton called a model environmental project) that will monitor everything from regional crises and territorial defense, to forests and oceans.[59]

Although the recent proliferation of environmental satellites should ensure that a military *monopolization* of satellite data does not take place, the trends outlined above are significant Adding weight to such issues is the fact that some influential conservative circles in the United States have been advocating the adoption of an aggressive American space policy.[60] In his last speech as head of the National Space Council, former Vice President Dan Quayle made a provocative bid for American "space control" including an "ability to deny the military use of space to future enemies."[61] Given the extent to which high-resolution imagery is increasingly available from civilian systems, such a policy of denial would presumably include these satellites as well.

Conclusion

The scope and complexity of long-term environmental monitoring necessitates the harnessing together of vast amounts of energy, expertise, and resources. In light of these circumstances, environmentalists may be tempted to embrace military institutions, where energy, expertise, and resources are plentiful. U.S. military satellite reconnaissance systems clearly offer unique advantages to the environmental community that are currently not widely available. Lending justification to such reorientation is the argument that military expertise is directed to more positive ends.

But for environmentalists, these are shallow benefits that mask the extent to which military and environmental institutions are fundamentally incompatible. To coordinate and assess the complex data needed to understand accurately ecological change, openness and objectivity are paramount. Yet intelligence organizations are among the most secretive of all institutions. Such secrecy could, in fact, nullify the benefits alluded to above. Environmentalists need assurance that no gaps in data exist, or that imagery that is provided by the NRO has not been altered for "national security reasons." Yet such assurances may never be obtained with complete certainty. Furthermore, environmental research is inherently global in scope. Consequently, ecological governance and environmental monitoring will require significant international cooperation. However, military institutions like the NRO may be reluctant to share classified data outside of a select predetermined circle of national researchers.

Environmental protection and military institutions may be incompatible in a more fundamental way. As many of the critics of the environmental security debate point out, the magnitude of environmental problems confronting humanity call into question the modern states system itself. By providing funding and programs to the military and intelligence organizations, environmentalists may, in fact, be serving to reinforce the very structures that must be transcended. The irony is that the turn to military organizations comes precisely at the time that civilian expertise on environmental matters is gathering momentum. Many sophisticated environmental satellites have been developed, while others await launch. Complex terrestrial infrastructures designed to coordinate and distribute the vast amount of data generated are in the embryonic stages of development. Most importantly, these systems are dedicated to the environment, while military systems have only tangential relevance. Environmentalists should be wary of incorporating the formidable military juggernaut into the enterprise of environmental rescue, where long-term goals may be pushed aside for the short-term demands of national security.

––––––––––––––––––––––––––– NOTES –––––––––––––––––––––––––––

1. Glenn Zorpette, "Sensing Climate Change," *IEEE Spectrum* (July, 1993), p. 20.

2. Stephen Walt, "The Renaissance of Security Studies," *International Studies Quarterly* vol. 35 (1991), pp. 211–39.

3. Jessica T. Mathews, "Redefining Security" *Foreign Affairs* vol. 68 (1989), pp. 162–77; and Theodore Sorenson, "Rethinking National Security," *Foreign Affairs* vol. 69 (1990), pp. 1–18.

4. Thomas Homer-Dixon, "Environmental Scarcities and Violent Conflict: Evidence From Cases," *International Security* vol. 19 (1994), pp. 5–40.

5. Peter Gleick, "Environment and Security: The Clear Connections," *The Bulletin of the Atomic Scientists* vol. 47 (1991), p. 19.

6. Daniel Deudney, "Environment and Security: Muddled Thinking," *The Bulletin of the Atomic Scientists* vol. 47 (1991), p. 23.

7. Trudy E. Bell, "Jobs at Risk," *IEEE Spectrum* (August 1993).

8. Daniel Deudney, "Environment and Security: Muddled Thinking," op. cit., p. 26.

9. Simon Dalby, "Ecopolitical Discourse: 'Environmental Security' and Political Geography," *Progress in Human Geography* vol. 16 (1992), p. 516.

10. A. H. Westing, "The Environment Component of Comprehensive Security," *Bulletin of Peace Proposals* vol. 20 (1989), pp. 129–34.

11. Simon Dalby, "Ecopolitical Discourse: 'Environmental Security' and Political Geography," op. cit., p. 513.

12. Daniel Deudney, "Environment and Security: Muddled Thinking," op. cit., p. 28.

13. Ken Conca, "In the Name of Sustainability: Peace Studies and Environmental Discourse," Presented at the Conference on Environmental Change and Security, Vancouver, 1993, pp. 9–10.

14. Russell J. Bruemmer, "Intelligence Community Reorganization: Declining the Invitation to Struggle," *The Yale Law Journal* vol. 101 (1992), pp. 867–77; and David L. Boren, "The Winds of Change at the CIA," *The Yale Law Journal* vol. 101 (1992), pp. 853–65.

15. Stansfield Turner as cited in Russell J. Bruemmer, "Intelligence Community Reorganization: Declining the Invitation to Struggle," op. cit., p. 870; see also Amy Borrus, "Should the CIA Start Spying for Corporate America?" *Business Week* (October 14, 1991).

16. William Burrows, *Deep Black: Space Espionage and National Security* (New York: Berkeley Books, 1986).

17. Comparisons have been made with NASA's Hubble Space Telescope, which is based on similar technology as the KH-11. If the Hubble were to

turn its lens toward the earth, it would have a ground resolution of 7.16 centimetres. Images received clandestinely by various media outlets over the years substantiate such a high capability level (Adams 1986, 49; Keydel 1990/1991, 55).

18. Jon Trux, "Desert Storm: A Space Age War," *New Scientist* (July 1991); and John Adams, "Peacekeeping by Technical Means," *IEEE Spectrum* (July 1986).

19. Craig Covault, "Recon Satellites Lead Allied Intelligence Effort," *Aviation Week and Space Technology* (February 1991).

20. Eric Schmitt, "Spy-Satellite Unit Faces a New Life in Daylight," *New York Times*, 3 November 1992.

21. Jon Trux, "Desert Storm: A Space Age War," op. cit., p. 30.

22. Of course, specific targets of satellite reconnaissance operations are highly classified. Nevertheless, a reasonable picture of such targets can be gained through a combination of deduction and analysis of the scant bits of information that are periodically filtered into public circles.

23. Jeffrey T. Richelson, *America's Secret Eyes in Space: The U.S. Keyhole Spy Satellite Program* (New York: Harper and Row, 1990), p. 245.

24. David C. Morrison, "Look Up—You're Being Imaged!" *National Journal* vol. 6 (1991), p. 1368.

25. Ironically, the U.S. military-industrial complex itself is now contributing to this problem. Facing commercial competition from Russian satellites and budgetary cutbacks, the U.S. private corporations that manufacture satellite systems have lobbied to undo Cold War government regulations prohibiting them from selling sophisticated systems and imagery (see Charles, 1992, 5; Lenorowitz 1993, 70).

26. William B. Scott, "High Demand Stretches NRO Intelligence Budget," *Aviation Week and Space Technology* (February 1993).

27. Ray Harris and Roman Krawec, "Some Current International and National Earth Observation Data Policies," *Space Policy* vol. 9 (1993), pp. 273–85.

28. For an overview of these different programs, see "Special Issue on ISY: Mission to Planet Earth," *IEEE Technology and Society Magazine* (Spring 1992).

29. Gary Taubes, "Earth Scientists Look NASA's Gift Horse in the Mouth," *Science* (February 1993).

30. James R. Asker, "NASA Reveals Scaled Back Plan for Six EOS Spacecraft," *Aviation Week and Space Technology* (March 1992).

31. Nahum D. Gershon and C. Grant Miller, "Dealing with the Data Deluge," *IEEE Spectrum* (July 1993).

32. Ray Harris and Roman Krawec, "Some Current International and National Earth Observation Data Policies," op. cit.

33. Keith Hilton, "Spaceship Earth," *Geographical Magazine Supplement* (March 1991).

34. Al Gore, "A New Initiative to Save the Planet," *Scientific American* (April 1990), p. 124.

35. Sam Nunn, "Harness Defense to Save the Environment," *Aviation Week and Space Technology* (July 1990), p. 7.

36. Telephone Interview, U.S. Naval Research Laboratory, 5 October 1993.

37. Vincent Kiernan, "Proposal Would Widen Access to Data from U.S. Spy Satellite," *Defense News* (August, 1993), p. 13; and Traci Watson, "Researchers See Obstacles in Using Spy Data," *Nature* (July 1992), p. 178.

38. William Broad, "Spy Data Now Open for Studies of Climate," *New York Times*, 23 June 1992.

39. See William J. Broad, "U.S. Will Deploy Its Spy Satellites on Nature Mission," *New York Times*, 27 November 1995.

40. Jeffrey T. Richelson, *America's Secret Eyes in Space: The U.S. Keyhole Spy Satellite Program*, op. cit., p. 270.

41. Jeffrey T. Richelson, *America's Secret Eyes in Space: The U.S. Keyhole Spy Satellite Program*, op. cit., p. 270.

42. William Broad, "Spy Data," op. cit.

43. Traci Watson, "Researchers See Obstacles in Using Spy Data," *Nature* (July 1992), p. 178.

44. Traci Watson, "Researchers See Obstacles in Using Spy Data," op. cit.

45. Daniel Deudney, "The Case against Linking Environmental Degradation and National Security," *Millennium: Journal of International Studies* vol. 19 (1990), pp. 464–65.

46. William Burrows, *Deep Black: Space Espionage and National Security*, op. cit., p. x.

47. William Burrows, *Deep Black: Space Espionage and National Security*, op. cit., p. viii.

48. Jeffrey T. Richelson, *America's Secret Eyes in Space: The U.S. Keyhole Spy Satellite Program*, op. cit., p. 111.

49. Jeffrey T. Richelson, *America's Secret Eyes in Space: The U.S. Keyhole Spy Satellite Program*, op. cit., p. 269.

50. As cited in Schmitt, "Spy-Satellite Unit," op. cit.

51. By fusing commercial imagery with digital terrain elevation data (DTED) simulations of approaches to specific targets can be made, enabling pilots to view the area from a number of different angles and trajectories prior to the actual bombing mission. The program can be stopped at any point in the simulated bombing run, enabling pilots to get a 360-degree view of potentially hostile force positions. See "SPOT Helped Bombs Find Iraqi Targets," *Military Space* (29 July 1991), p. 1; and "Army Using Satellite Imagery for Bosnia," *Military Space* (8 March 1993), p. 5.

52. See "Army Using Satellite Imagery for Bosnia," *Military Space* (8 March 1993); "Civil Images May Have Fuelled Third World Wars," *Military Space* (31 July 1989).

53. "Army Using Satellite Imagery for Bosnia," op. cit.

54. "Civil Images May Have Fuelled Third World Wars," op. cit.

55. Eliot Marshall, "Space Cameras and Security Risks," *Science* (March 1989), p. 472.

56. *Assessing the Potential for Civil-Military Integration: Technologies, Processes, and Practices*, United States Office of Technology Assessment Report ISS-611 (September 1994).

57. See Joanne Gabrynowiez, "The Promise and Problems of the Land Remote Sensing Act," *Space Policy* 9 (1993), pp. 319–28.

58. See *Civilian Satellite Remote Sensing: A Strategic Approach*, United States Office of Technology Assessment Report ISS-607 (September 1994), especially chapter 3, "Planning for Future Remote Sensing Systems."

59. "Brazil Signs $1.4 Billion Satellite Contract with U. S. Companies," Associated Press/News and Observer Publishing Co., 1 June 1995.

60. William Burrows, "U.S. Space Grab," *New York Times*, 27 March 1993.

61. William Broad, "Quayle Urges Developing Space Weapons for Conventional Wars," *New York Times*, 15 January 1993.

Conclusion

12

Conclusion
Settling Contested Grounds
Richard A. Matthew

This volume was prepared to provide students, researchers, and policymakers with a comprehensive and up-to-date overview of a growing subfield of international politics: environmental security and conflict. As discussed in chapter 1, during the 1970s environmental issues were placed on the global agenda. The writings of this period often were characterized by a sense of urgency and outrage—we were biting the hand that fed us, and it was about to close into a fist. Scholars in the 1980s and early 1990s tended to adopt a more objective, elaborate, and less confrontational tone. They sought to clarify the complex relationships between environmental change and human security and welfare in a manner that took into account both disputes within the scientific community and the multitude of constructive and destructive political and economic forces operating in the global arena.

In the late 1990s environmentalism has experienced a sort of conceptual crisis. Among the converted, it has fragmented into at least four perspectives: deep ecology, social justice ecology, technological optimism, and conservationism. Each group has sifted through the scientific evidence, established priorities, assessed the utility of *in situ* political, economic, and ethical systems, and developed prescriptions. While the fragmentation has generated a rich and fruitful debate, it has left outsiders confused. What exactly is pollution or environmental change? Is our world on the verge of collapse, or do we have a basket of concerns that should be weighed against other concerns and addressed when and if they cross some threshold?

In this context of debate, eco-skepticism has a simple appeal. After all, humankind lived through the peak of the last ice age armed with very rudimentary tools; surely it can cope with a few degrees of global warming or the overexploitation of fisheries. Thus, in addition to considering research on the implications of environmental change, observers might be wise to ask, Who stands to benefit from this issue area? Is environmental change truly the security issue of the late twentieth century, as some have claimed?

Contested Grounds summarizes many of the findings of the past twenty years. Readers hoping for a compendium of straightforward causal relationships and clear policy recommendations may feel disappointed. In some cases, significant work has been done to reinforce positions that seem obvious—some forms of environmental change threaten human welfare and may trigger or reinforce some forms of violence. In other cases, the debate has shaded into a familiar set of controversies about the complexity of interstate conflict, North-South disparities, the status of the military in world affairs, the viability of the state, and the countervailing pressures for integration and fragmentation in international relations.

In the twilight of the twentieth century, underscoring the weaknesses, failures, and methodological and epistemological problems apparent in the social sciences has become the blood sport of results-oriented intellectuals and politicians. This attitude toward scholarly research and debate is tiresome and uninformed. It is true that *Contested Grounds* does not deliver a definitive statement on the problem of environmental security and conflict. But it contributes to the evolving—and vital—project of rethinking the relationship between nature and civilization.

It is important to situate the work presented in this volume in the broader—and interrelated—contexts of security issues, environmentalism, and international relations. *Contested Grounds* concludes what Mark Levy has described as the "second wave" of research and debate on environmental security and conflict.[1] As such it serves an important role in the more general process of rethinking world politics in the post–Cold War era.

Implications for Security Issues

Several charges have been made against the research on environmental security and conflict that has been conducted over the past few years. A common criticism, reflected in the second Deudney article in this volume, is that much research has been constrained by flawed methodological frameworks. Too much attention has been given to demonstrating the obvious: that linkages exist between environmental change and human security and welfare. Too little has been given to ascribing a clear weight to this relationship, investigating the precise conditions under which it is preponderant, and placing it in a satisfactory historical context. These are valid concerns, and it is certainly through this type of critique that progress is made in the social sciences. But in some measure, these criticisms undervalue the contributions of this research.

First, it is largely due to this research that environmentalism has penetrated the security community of academics and policymakers. The work of Homer-Dixon and others has placed environmental change on a closely guarded agenda. The cautionary statements of Dalby, Deudney, and others have helped to ensure that competing perspectives are taken into consideration. As Butts notes in his chapter in this volume, the U.S. military's mandate includes environmental security—and this is because of the attention given to it over the past fifteen years by scholars willing to explore what was once a marginal area in the field.

Second, it is desirable that much research has highlighted the fact that environmental change plays a variable role in the complex of forces that produce insecurity and violence. For forty-five years security thinking referred to a single baseline: containing Soviet expansion. Recent scholarship corrects the flaws inherent in this perspective by underscoring the multiple and interactive sources of insecurity and the importance of developing a sophisticated array of prevention and response capabilities.

For the practitioner the message is clear—there are no silver bullets. Very general theoretical models and theories do not explain everything. At best, they offer fascinating perspectives and provide basic principles to guide policy formulation. But actions need to be informed by first-rate local expertise and knowledge. We need, in short, to avoid the errors associated with the uncritical acceptance of very general analysis.

Another common criticism focuses on the lack of conceptual rigor associated with the "second wave" of scholarship. What, exactly, does environmental security mean? Clearly, it means different things to different people—largely because people are threatened, or feel threatened, in a multitude of ways. Some are threatened by what Bill McKibben has called the "end of nature"; others by polluted water, the spread of various diseases, or a lack of arable land; and still others by the disruption of economic activities, the movement of people into their territories, or the threat of war. What constitutes security, and what threatens it, vary and research on environmental security reflects this.

A third criticism laments the gap between theory and practice, or the slow pace at which new ends are being institutionalized. A prevalent sentiment is that we have had twenty-five years of talk, but precious little policy action. This is a valid criticism. But the research of recent years is part of the critical mass that will help to pick up the pace. The vestiges of Cold War thinking are insistent but cannot last forever. The nuclear generation is giving way to the ecological generation, which incorporates and expands upon the fears and aspirations of its predecessors. And the close linkages between environmentalism and every other aspect of human life can gradually resolve into policies that are multilateral, compatible, and progressive.

Evidence that such changes are occurring can be found in activities such as the Woodrow Wilson Center's multiyear Environmental Change and Security Project. This project brings together representatives from various government agencies, nongovernmental organizations, and the academic world for regular discussions about what environmental security means, what they can do to promote it, and how their efforts might be coordinated.

The process of coming to terms with security and conflict at the close of the twentieth century poses a serious challenge to a world accustomed to thinking in terms of military aggression and the integrity of the sovereign nation-state. But as Mark Zacher, John Mueller, and others have argued, this era is almost over.[2] In today's interdependent world, the most pressing threats to human security require a broadening of the concept and the development of new tools. The chapters in *Contested Grounds* are a valuable part of this ongoing process of redefinition.

Implications for Environmentalism

A persistent—but inaccurate—strand of contemporary thought continues to depict environmentalism as a single-issue movement. Such a caricature sets up an easy target—environmentalists are zealous lobbyists, unsympathetic to job security, convenience, technological innovation, scientific research, economic growth, the sanctity of human life and a host of other real-world issues and achievements. In fact, environmentalism has evolved rapidly as a comprehensive ideology that incorporates various beliefs, values, practices, and institutions with the objective of maximizing human welfare and potential in a way that does not destroy the life-support system upon which they depend.

In the past twenty-five years, three areas have gained a place on national and global agendas and stimulated far-reaching research and debate—environmental security, sustainable development, and environmental justice. The work done on environmental security and conflict has been integral to the process of broadening the politics of environmentalism and ensuring that it takes into consideration the full range of human needs, fears, and desires. It is because of this sort of research that the baby boom generation's culture of narcissism and its romance with the past—explicit in characterizations of environmentalism as a single-issue movement—are giving way to ideas and practices more appropriate to the needs of the day.

Contested Grounds contributes to the broadening of environmental politics as well as the broadening of security affairs. There are individuals in both communities who are wary of such efforts—who fear that the clarity of security thinking, which enables military readiness, will be undermined, or that the revolutionary potential of environmentalism will be sacrificed as it moves into the mainstream. While such resistance is often regrettable and obstructionist, it would be unwise not to take these concerns seriously. The world is still composed of formally sovereign states, many of which pose traditional security threats, and the capacity to respond effectively must be maintained. Moreover, part of the appeal and value of environmentalism lies in the general and comprehensive challenge it poses to contemporary industrial culture and the ideologies that sustain it. But neither traditional state-centric security thinking or the extreme forms of environmentalism are likely to be adequate in mobilizing support and resources, or responding to the diverse needs of humankind and its life-support system in the years ahead. Both will have to find ways of reconciling established practices with emerging ones, of forging new patterns of behavior, and of developing new goals. This can only be achieved through discussion, research, and experimentation. *Contested Grounds* brings together two erstwhile independent issue areas in a constructive and sensitive manner designed to stimulate the activities required for reaching a useful consensus and program of action.

With this in mind, it is important to reflect more deeply on the manner in which environmental concerns are being integrated into the traditional categories of international politics: security, trade and development, and justice. The general project of negotiating a more satisfactory relationship between nature and civilization requires more than simply the greening of traditional categories of behavior. The complex relationships between security, economics, and ethics require further investigation than they have received to date.

Today there is a trend toward dividing environmentalism into somewhat discrete issue sets, which then develop along more or less independent trajectories. At times, it will be crucial to find ways of looking at the big picture, so that we can try to reconcile competing or contradictory policy initiatives. Or, put another way, institutionalizing environmentalism via existing structures is necessary and desirable, but poses certain risks. In particular, there is the risk that we will tend to treat symptoms, respond to crises, and remain reactive even in those cases where a proactive approach would be more cost efficient and effective.

The problem is, for example, that a proactive approach to reducing environmentally generated conflict might require aggressive economic measures. If the researchers exploring environmental change and conflict, and the agencies establishing guidelines for responding to this, are too narrowly focused, their efforts may well be suboptimal. To avoid this, a permanent dialogue between specialists in the areas of environmental security, sustainable development, and environmental ethics should be encouraged. Moreover, some thinkers must continue to work at the most general level, informed by the very fundamental problem of the relationship between civilization and nature.

Implications for International Politics

In the academic discipline of international politics, two debates have emerged in recent years that inform much of the teaching and research that takes place. The first is between neoliberal institutionalists and their critics, especially realists and neorealists. The former focus on the forms of interdependence, the growing menu of shared interests, the proliferation of diverse actors, and the possibilities of cooperation, principally through regimes, that characterize contemporary international politics. The latter emphasize the anarchic nature of the international system, the extent to which interests conflict, the continuing predominance of the state, and the many obstacles—such as the relative gains problem, cultural differences, and vast inequalities of power and knowledge—to cooperation. This debate has a long history in the field. What is significant today is that, with the end of the Cold War and the indisputable importance of trade and development issues for all world actors, the neoliberal position appears to be gaining the upper hand over realism, which was the preferred worldview of the majority of Western academics during the four decades following World War II.

The second debate has served to rotate the principal axis of confrontation in international politics from East-West to North-South. In very general terms, this debate is characterized by disagreements on the applicability of the North's development experience to the South, the sources of inequality between the two "regions," and how the relationship between the two can and should evolve in the years ahead. Although the Cold War was marked by periods of detente, the clear objective throughout this period was to contain, and eventually eliminate, the threat of Communist expansion posed by the Soviet Union. Insofar as North-South relations are concerned, however, there is no clear objective. The South is deeply fragmented and highly diverse, and the North's interests in these fragments vary from state to state—and year to year.

The new environmental politics, one dimension of which is presented in *Contested Grounds*, cuts into these debates in a variety of ways. First, scientific evidence compellingly demonstrates that many environmental issues (such as ozone depletion, deforestation, global warming, the depletion of fisheries, and many forms of air and water pollution) are transnational or even global in scope. These types of environmental change can stimulate migration, exacerbate intrastate and interstate tensions, and create the conditions for conflict.

To address these issues effectively is likely to require various forms of multilateral cooperation. Thus, at first blush, the neoliberal institutionalist perspective appears more compatible with environmental politics than its principal alternatives. In this vein, scholars such as Robert Keohane, Marc Levy, Peter Haas, and Oran Young have analyzed the formation of environmental regimes. The promise of environmental security regimes is one worth exploring.

Neoliberal institutionalism has somewhat less to say, however, about how to deal with situations in which multilateral cooperation is desirable, but mutual interests have yet to be articulated or appear not to exist. Since this situation characterizes many environmental issues, it deserves considerable attention. How do we maximize environmental security when a state's activities constitute an environmental security threat, but it regards them as essential to its economic well-being? It is very likely that all states might be described in these terms. Brazil and Indonesia cut down rainforests, Russia sends radioactive contamination over its borders, Spain depletes Atlantic fisheries, China burns vast quantities of low-grade coal, and the economic web centered in the United States sustains environmentally unsound practices in many regions of the world.

This problem is linked in important ways to the second major debate in the field over North-South relations. The priorities of these two regions, in very general terms, are often at odds and environmental politics frequently has been derailed by strident blamecasting and efforts to compel or persuade the other side to take the initiative and bear the costs of environmental rescue. Should we act to control Southern population growth or curtail Northern consumption practices? Modernize the South through environmentally sound technologies or revise unsound technologies embedded in existing structures in the North? Redistribute wealth or apply force? Interests conflict and strong motivations to maximize environmental security do not yet exist. Politicians face incessant demands to assign scarce resources and expertise to a host of pressing needs. Leadership on this issue has yet to emerge.

The lens of environmentalism, however, does have the potential to help us move beyond this impasse, at least conceptually, insofar as it underscores the various levels of interdependence that exist within and between issue areas. We cannot expect environmental change to generate clearly defined mutual interests on a regular basis. But as we come to appreciate the complex relationship between nature and civilization in its multiple forms, rational policy positions are likely to emerge. For example, controlling population growth through contraception may reduce stress on certain ecosystems. But population growth is linked to a range of variables, including poverty and the inappropriate use of technologies which also place stress on ecosystems. A rational, fair, and effective policy will have to act upon several of the sources of population growth. Moreover, the impact of population growth on the environment is also related to issues such as access to resources. This, too, needs to be considered if a viable policy is to be formulated and successfully implemented.

Looking Ahead

As noted in chapter 1, at the most general level, an environmentally secure world would have three basic characteristics. First, it would exploit environmental goods at a sustainable rate. Second, it would allow some fair form of universal access to these goods. Third, it would manage crises and conflicts arising from inevitable instances of scarcity.

The dilemma we face is that while, from one perspective, everyone may be assumed to have an interest in the form of environmental security envisioned above, diverse particular perspectives on the causes of insecurity, the entitlements of different groups, the hierarchy of threats, and the form of fair and practical solutions make it difficult to manage the supply and consumption of environmental security. Power differentials further complicate the process, as they create different incentives for competitive extraction and long-term payoffs.

Ironically, it is not hard to imagine, in general terms, what should be done:

- reduce the consumption of environmental goods in the North

- improve efficiency in the North so that reductions in the consumption of environmental goods do not translate into a huge drop in the standard of living

- improve the access to environmental goods of the South and within the South

- as access improves, eliminate unsustainable practices in the South

- restrain population growth

- cooperatively manage the conflicts that will inevitably arise due to the differences in local and global imperatives, short-

term and long-term interests, and subjective and objective
values and beliefs, that, together with disruptive political
forces, will adversely affect efforts to provide and maintain
global environmental security.

Conceptually, significant progress has been made in defining the
common objectives and standards, appreciating the inevitable conflicts between short-term interests, and envisioning ways of reconciling short-term and long-term payoffs that are necessary to move the
project forward. In some measure, the world knows what has to be
done and has sketched viable routes toward the achievement of environmental security. But compelling intuitions, well-intended end-
games and general policy principles have not proven sufficient to mobilize the large-scale transitions that are both necessary and
desirable.

Some observers have contended that environmental security
thinking has stalled at a level of generality that is inoperable. Others
have noted that all environmental policy has reached a sort of cross-
roads where the complexity of the problems and the huge menu of
options that have been identified have confused and disoriented policymakers. In both cases there is a demand for concrete, low-cost policies that are appropriate to different contexts and that are nonetheless complementary.

In the absence of such policies, predictions of the future tend to
be bleak. Some are committed to the belief that an environmental
crisis of huge proportions will be required to generate the type of
multilateral cooperation and forward-looking thinking that is required to achieve environmental security. Waiting for an environmental crisis, however, is not only pathetic but could prove futile:
should such a crisis occur the incentives to compete aggressively
might increase. Others imagine a growing gap between North and
South that will create the conditions for violent confrontation—given
the destructive capability of modern weaponry and its increasing accessibility, this is a dismal scenario. Some optimists predict a gener-
ational shift in policy priorities as the cold warriors still in power are
replaced by rainbow warriors more aware of the need to address environmental issues aggressively and cooperatively. And a number of
pessimists envision a world in which we simply adapt to environmental change, accepting the steady erosion of freedom, welfare, and
security that it is likely to entail.

Is a more optimistic and action-oriented program conceivable?
The would-be consumers of environmental security must mobilize to

compel powerholders to promote their interests. The incentives that make competitive extraction more immediately desirable than cooperative management must be revised. And practices that support discounting the future must be challenged. This three-pronged approach to environmental security must also be sensitive to local and global differentials in cause and effect.

Conclusions

In this volume the concept of environmental security has been placed into a broad historical context, defined from several overlapping but distinct perspectives, and examined through case studies that span North and South, East and West. Clearly, there is no silver bullet that will address the threats posed by environmental change. But there are steps that can be taken—by researchers, policymakers and activists—to restore the complex equilibrium between nature and civilization that must be restored and protected if human welfare, security, and freedom are to be preserved. The problem lies in reconciling the many sources of disequilibrium—most of which are rooted in narrowly defined human aspirations and fears.

Solving this problem now has a large pool of research to support it. It is not yet complete, but it is moving in a fruitful direction. *Contested Grounds* summarizes where we stand today, and suggests what we must do in the years ahead.

-------------------- NOTES --------------------

1. Marc A. Levy, "Time for a Third Wave of Environment and Security Scholarship?" *The Environmental Change and Security Project Report* Issue 1 (Spring 1995), pp. 44–46.

2. Mark Zacher, "The Decaying Pillars of the Westphalian Temple: Implications for International Order and Governance," James N. Rosenau and Ernst-Otto Czempiel, eds., *Governance without Government* (Cambridge, U.K.: Cambridge University Press, 1992); John Mueller, *Retreat from Doomsday: The Obsolescence of Major War* (New York: Basic Books, 1989).

About the Contributors

KENT HUGHES BUTTS is professor of Political Military Strategy in the Center for Strategic Leadership at the U.S. Army War College, and holds the George C. Marshall Chair of Military Studies. A former John M. Olin Fellow in National Security at the Center for International Affairs, Harvard University, he is the War College Environmental Course Professor and author of numerous articles, monographs, and book chapters as well as a book on geography and security.

SIMON DALBY is associate professor at Carleton University in Ottawa where he teaches geography and international affairs. He is the author of *Creating the Second Cold War* (Guilford, 1990). He recently coedited *The Geopolitics Reader* (Routledge, 1998) and *Rethinking Geopolitics* (Routledge, 1998).

RONALD J. DEIBERT is assistant professor in the Department of Political Science at the University of Toronto, specializing in technology, media, and world politics. He is the author of *Parchment, Printing, and Hypermedia: Communication in World Order Transformation* (Columbia University Press, 1997).

DANIEL DEUDNEY is assistant professor of political science at Johns Hopkins University. He has written extensively on international relations theory, environmental politics, and geopolitics.

MICHEL FRÉDÉRICK holds degrees in public international law (LLM) and political science (PhD). He works in Quebec City as a consultant in the field of environmental law and policy. He also lectures in international law at Laval University Law School.

JACK A. GOLDSTONE is professor of sociology at the University of California, Davis. His articles on population history and political conflict have appeared in *Population Studies*, *World Politics*, the *Oxford Companion to Politics of the World*, and many other books and journals. He is the author of *Revolutions of the Late Twentieth Century* and *Revolution and Rebellion in the Early Modern World*.

THOMAS F. HOMER-DIXON is director of the Peace and Conflict Studies Program and assistant professor in the Department of Political Science at the University of Toronto. He was the principal investigator of the Project on Environmental Scarcity, State Capacity, and Civil Violence, and director of the Fast Track Project on Environment, Population and Security, sponsored by the program and the American Association for the Advancement of Science (AAAS) in Washington. From 1990 to 1993 he was codirector of an international research project on Environmental Change and Acute Conflict also organized by the Peace and Conflict Studies Program and the Academy. He has published on a range of topics from arms control to environmental policy to the philosophy of the social sciences and has lectured widely in North America and overseas.

MIRIAM R. LOWI is assistant professor of political science at The College of New Jersey. She has written extensively on conflict in international river basins in the Middle East and South Asia. Her book *Water and Power: the Politics of a Scarce Resource in the Jordan River Basin* was published by Cambridge University Press in 1993, and an updated version appeared in 1995. Her current research focuses on oil and the politics of development in Algeria, Iran, and Indonesia.

RICHARD A. MATTHEW is assistant professor of international relations in the School of Social Ecology, University of California at Irvine. He has published articles on international relations theory, ethics in international affairs, and environmental issues. He is currently completing a book entitled *Shared Fate: Ethics, World Politics and the Environment*.

ERIC STERN is a research fellow and assistant professor of political science at Stockholm University. He has written extensively on security topics.

Index

acid rain, 63, 105
Afghanistan, 136, 206
Africa, 34, 44–50, 161, 197
African National Congress, 71
Agarwal, Anil, 165
Agenda, 5, 21
Agent Orange, 136
Agnew, John, 180n.12
AIDS, 133
air conditioning, 36
Algeria, 165, 206
Alliance of Small Island States (AOSIS), 131
Almond, Gabriel, 43
Amazon: region, 49–50, 168
American Academy of the Arts and Sciences, 189
Amin, Idi, 213
Amnesty International, 133
Anglo-Islandic Cod War, 70
anthropomorphism, 7
Antarctica, 167
anthropological naturalism, 29

apartheid, 79–80
Apollo Project, 142
Arafat, Yasir, 238
Argentina, 207
Aristotle, 26, 31, 34–35
Aristride, Jean-Bertrand, 81
Arnold, David, 53n.25
Assam, 73–74
Aswan High Dam, 207
Asyut, Egypt, 77
authoritarianism, 84–85, 132, 198, 210–211

Bafing River, 66
Bangladesh, 72–74
Barraclough, Geoffrey, 31
Benedick, Richard, 183n.41
Bhopal, India, 136
bipolarity, 92
Black Death, 44
Bodin, Jean, 35
Booth, Ken, 152n.58
Bosnia, 137

Botero, 35
Brahmaputra, River, 72–74
Brandt Commission, 133
Braudel, Fernand, 31
Brazil, 85, 200, 207, 212; Amazon
    Survey System, 238
Britain, 45
Brodie, Bernard, 212
Brown, George, 277
Brown, Lester, 1, 3, 130, 188
Brunei, 208
Brundtland Commission, 95, 133
Brundtland, Gro Harlem, 5
Burke, Edmund, 201
Burkina Faso, 75
Buzan, Barry, 99, 129, 131, 139, 144,
    184n.61, 215n.14

Cairo: population conference, 77, 92,
    111
Calcutta, 74
Caldwell, Lynton, 3, 149n.18
Cambodia, 137, 213
Cape Province, 80
capitalism, 45–46, 163–165, 204
Carlsson, Ingvar, 133
Carson, Rachel, 3
Carter Doctrine, 9
Carter, Jimmy, 198
Central Intelligence Agency (CIA),
    124, 273
CFC's, 166–167, 175, 199
Chernobyl, 136
Chesapeake Bay, 121
China, 70, 136, 171, 175, 210, 248–
    266; regional conflicts, 252; Man-
    chu dynasty, 250; Ming dynasty,
    250; one-child family policy, 248;
    special economic zones, 254–257;
    Tiananmen Square Uprising,
    253–254, 256, 260
Ciskei, 79–80
Clay, Jason, 197
climate, 34–36; change, 4, 10, 13, 63,
    95, 101, 130–134, 137, 141–142,

161, 164–165; effects of, 26, 30,
    34–36; historical impacts, 42
Clinton, Bill, 134, 283
Clinton Administration, 156, 215n.8
coercive conservation, 169–171
Cold War, 1, 2, 91, 109, 112–115,
    124, 156, 158–160, 188, 268, 275–
    277, 281–282, 293–294, 297
Commission on Global Governance,
    133, 156–157
Commission on Sustainable Devel-
    opment, 105
Committee on Earth Observation
    Satellites (CEOS), 275
Commoner, Barry, 3
commons: global, 167–169, 209–
    210; tragedy of, 3
Conca, Ken, 152n.55, 182n.44,
    183n.54, 189, 195
Confidence and Security Building
    Measures, 137
Connelly, Mathew, 10, 11
conservationism, 7–8, 97, 292
cornucopians, 210–211
cosmic naturalism, 28
Crosby, Alfred, 31, 47
Council on Foreign Relations, 278
Cultural Revolution, 258

Dalby, Simon, 145, 147n.4, 154n.75,
    189
Darwin, Charles, 27, 28, 40–42
Debelko, Geoffrey, 179n.5, 179n.9
debt-for-nature exchanges, 105
Declaration of Principles on Interim
    Self-Government Arrangements
    (OSLO I), 224
deep ecology, 6, 7, 13, 292
Defense Environmental Restoration
    Act (DERA), 122
Defense Environmental Restoration
    Program (DERP), 122
democratization, 12, 124, 256, 261,
    264
denatured social science, 30

Der Derian, James, 197
determinism, 32–33, 81–84
Deudney, Daniel, 52n.12, 138–140, 145
Dewey, John, 146
disease, 46–50, 192
drugs: war on, 134, 161, 195, 274
Duvalier regime, 81

Earth Observation International Coordination Group (EO-ICWG), 272
earth nationalism, 201–202
Earth Observing System (EOS), 274
East Pakistan, 72
ecocentricism, 8
ecological marginalization, 65–66
economic deprivation, 74–78
economic security, 9, 93, 130, 135, 292
ecoskepticism, 7, 8, 14, 292
ecotourism, 170
Egypt, 71, 77, 206
Ehrlich, Paul, 3
El Salvador, 70
end of nature, 294
Environmental Change and Security Project, 189, 294
environmental crisis, 2, 3, 6, 94, 300
environmental justice, 163, 295
environmental rescue, 18, 19, 267–288
environmental scarcity, 64, 75
environmentalism, 2–3, 201, 292–301
Ethiopia, 71
ethnic conflict, 63, 72–74, 114, 139
Euphrates, 207
Eurasia, 44
European expansion, 42–50
European Organization for the Exploitation of Meteorological Satellites (EUMETSAT), 276
European Space Agency (ESA), 276
European Union, 4, 145

exogeneity, 82

Faga, Martin, 281
Febvre, Lucian, 32
feudalism, 44
Finger, Matthias, 152n.55, 189
Food and Agriculture Organization (FAO), 4, 66, 96
foreign policy decision-making, 143–144
Foreign Assistance Act, 115
France, 136, 206
Friedman, Milton, 43
Friends of the Earth, 5
Fukuyama, Francis, 12
functionalism, 19, 29

Gaia Hypothesis, 28
Ganges, River, 72–74
geopolitics, 26, 29, 156, 159–178
geopiety, 201
George, Alexander, 143
Germany, 132, 205, 208
Gibbon, Edward, 37, 40
Glacken, Clarence, 52n.2
Gleick, Peter, 88n.17, 207, 270
Global Change and Remote Sensing Project, 278
global warming, 13, 131, 133, 164–166, 168–169, 175, 210, 278, 292, 297
globalism, 12, 97, 134, 138–140
Goeller, H. E., 206
golden mean, 34
Goodman, Sherri Wasserman, 110
Gorbachev, Mikhail, 263–264
Gore, Al, 150n.30, 277
Greeks, ancient, 34, 46
green diplomacy, 102
Greenpeace, 5, 6

Haas, Peter, 298
Haiti, 79–81
Hanford, Washington, 112
Hanseatic League, 45

Hardin, Garrett, 3
Hawley, T. W., 136
Heartland, 33
high politics, 141–142, 222
Highlands Water Project, 71
Hippocrates, 34, 37
Hitler, Adolf, 32, 42, 205
Hobhouse, Leonard, 41–42
Holland, 45
Homer-Dixon, Thomas, 86n.1, 137, 139, 157, 189, 212
Honduras, 70
Howard, Michael, 196
human rights, 7, 12
Huntington, Ellsworth, 32, 35–36
Hurrell, Andrew, 180n.17, 182n.44
Hussein, Saddam, 112, 191

India, 72–74, 171, 212
Indonesia, 85
intelligence community, 10, 11, 272
International Labor Organization (ILO), 4
International Maritime Organization (IMO), 4
international relations, field of, 25–26, 31
International Space Year (1992), 276
Iran, 208
Iran-Iraq War, 164
Iraq, 207, 208
Israel, 175, 207, 223–245; Green line, 229–232; Water Commission, 221–233; Occupied Territories, 233; UN Armistice Demarcation, 229
Israeli-Palestinian Interim Agreement on the West Bank and Gaza Strip (OSLO II), 238
Israeli-Palestinian Joint Water Committee, 233

James, William, 195
Japan, 70, 159, 162, 205, 208, 211

Jerusalem, 223–245
Jordan, 221–243
Jordan, River, 223–245

Kahn, Herman, 6, 8
Kaishek, Chaing, 253
Kakonen, Jyrki, 196
Kamarck, Andrew, 48
Kaplan, Robert, 10, 11, 181n.28, 189
Katzenstein, Peter, 132
Kennedy, John F., 142
Kennedy, Paul, 10, 11
Kenney, Ralph, 144
Keohane, Robert, 298
KGB, 132
Khaldun, Ibn, 37–38
Khmer Rouge, 213
Kidd, Benjamin, 41
Kjellen, Rudolf, 41
Klausner, Samuel, 30
Kristof, Ladis, 52n.5
Kroptkin, Peter, 41–42
Kuwait, 136, 191, 208

Lalung tribe, 73
landmines, 137
Lamb, H. H., 54n.33
leprosy, 48
Lesotho, 71
Levy, Marc, 140–142, 179n.6, 189, 293, 298
liberalism, 6, 12, 105, 132, 197–198, 200, 204
limits to growth thesis, 9
Lipschutz, Ronnie, 181n.34, 185n.79, 205
Little Ice Age, 36
Liverman, Diana, 86n.4
Lodgaard, Sverre, 106n.2, 137
Love Canal, 94
Lovins, Amory and Hunter, 164
low politics, 141–142, 224, 270
Luzon, 74

Machiavelli, Niccolo, 37

Mackinder, Halford, 33
Magat watershed, 74
malaria, 48
Malaysia, 68, 72
Mali, 66
Malthus, Thomas, 28
Manantali Dam, 66
Manila, 67
Marcos, President, 77–78
Maqarin Dam, 236–237
Markham, S. F., 35–36
martial virtue, 36–40
Marxism, 6, 33, 43, 187
Mathews, Jessica Tuchman, 10, 11, 130, 155–156, 189
Mauritania, 66–67
Mazrui, Ali, 49
McCormick, John, 3
McKibben, Bill, 294
McNeill, William, 31
Meadows, Donella, 3
Mekorot, 232–234
Merchant, Carolyn, 6
Mesoamerica, 42, 45
Mesopotamia, 42
Middle East, 42, 44–46, 113, 117, 223–245
Mies, Maria, 6
migration, 46–48, 72–81, 134, 161, 172, 252
mineral resources, 30, 70, 205–206; substitution, 206
Mische, Patricia, 106n.7, 188
Mission to Planet Earth, 276
Mongol invasions, 44–45
Montesquieu, 26, 30, 31, 34–35
Mueller, John, 294
multicausality, 83
Myers, Norman, 106n.5, 155, 183n.58, 189, 198

Narain, Sunita, 165
National Security Strategy of the United States, 2, 110, 113
nationalism, 189–190

natural-social science, 30
Nature Conservancy, 122
neoliberal institutionalism, 298
Neo-Malthusians, 210
neorealism, 297
New Zealand, 47
Nigeria, 85
nomads, 36–40
nonlinearity, 83
North Atlantic Treaty Organization (NATO), 4, 117–118, 139, 156, 159, 161–163, 172; Committee on the Challenges of Modern Society, 117; North Atlantic Cooperation Council, 117; North Atlantic Council, 116
nuclear: accidents, 94, 136, 191; weapons, 136, 191, 204–205, 209
Nunn, Sam, 109, 277–278

Oaxaca, Mexico, 64
oil crisis, 9, 93, 161
onochocerciasis, 48
OPEC, 9, 161
Organization for Economic Cooperation and Development (OECD), 4, 96
Organization for Security and Cooperation in Europe, 140
Organization of African Unity, 4
ozone depletion, 10, 63, 95, 97, 101, 105, 130–131, 141–142, 175, 199, 278, 297

Palestine, 223–245
Paraguay, 207
Parana River, 207
Paris Peace Conference (1919), 231
Partial Test Ban treaty, 135
peace studies, 128
Pearson, Karl, 41
Peluso, Nancy, 183n.54, 197
Peng, Li, 255
Persian Gulf War, 20, 70, 136, 164, 191, 208, 232, 274, 282

Peru, 84
Philippines, 67–68, 74, 77–78, 84;
    National Democratic Front, 77;
    New People's Army, 77
Pirages, Dennis, 51n.1, 184n.59
population: control, 171–172; crisis,
    247–266; displacement, 10, 71;
    growth, 3, 10, 19, 61, 72, 81, 113,
    134, 171, 227–228, 298–299; his-
    tory, 30, 46–49; pressures, 19, 64–
    67, 117
Port-au-Prince, 80–81
possibilism, 32
physiopolitics, 29
Prins, Gwyn, 178n.2, 185n.77, 189
protectionism, 93
psychology, 34

Quayle, Dan, 283

Rabin, Yizhak, 238
race, 29, 40–42, 66, 81
Ramphal, Shridath, 133
Raspail, Jean, 10
Ratzel, Friedreich, 32, 41
realism, 25–26, 31, 41, 91–108, 134–
    138, 297
regionalism: economic, 92; biore-
    gions, 202
Renner, Michael, 106n.7, 156, 189
Repetto, Robert, 74
resource capture, 65–66
Rome, ancient: Republican, 38; Im-
    perial, 40
Rongji, Zhu, 254–255
Russia, 209, 210
Rwanda, 139

Sadat, Anwar, 206
satellites: imagery, 279–280; recon-
    naissance, 268–288; Defense
    Communications System, 274;
    Defense Meteorological Support
    Program, 278; Keyhole KH, 273–
    288; LACROSSE, 273; NAVSTAR

Global Positioning System GPS,
    274
scarcity, 7, 9, 13; resource, 10, 16, 19;
    water, 19, 117, 223–245
schistosomiasis, 48
security studies, 26, 128, 157
Senegal, 66–67
Senegal River Valley, 66, 71
Shangkun, Yang, 255
Sharett, Moshe, 231
Shiva, Vandana, 6, 166
Simon, Julian, 6, 8
Soccor War, 70
social Darwinism, 28, 40–42
social ingenuity, 69
social justice ecology, 7, 14, 292
social-social science, 30
soil: degradation, 63–64, 68, 80; fer-
    tility, 36–40, 174; shortage, 249–
    266
Somalia, 118, 139
Soroos, Marvin, 185n.77
South Africa, 71, 79–80
sovereign immunity, 123
sovereignty, 91, 93, 99, 145, 160,
    167–169, 176, 188, 198
Soviet Union, 91–93, 99, 112, 116–
    118, 136, 142, 191, 206, 268, 275,
    293, 297
Sparta, 38
Spencer, Herbert, 40–41
Sprout, Harold and Margaret,
    51n.1, 53n.19
state: capacity, 84–85, 197–198,
    200; collapse, 16, 84–85, 212–
    213; system, 132–133, 140, 145,
    196, 204–213; territorial integ-
    rity, 94, 98–99
Stern, Eric, 153n.61, 178n.2, 198
Strategic Arms Limitation Talks
    (SALT), 280
structural adjustment, 174
Sundelius, Bengt, 130
sustainable development, 10, 95,
    102, 114, 124, 168, 295–296

synthetic aperture radar (SAR), 273

Tahal: Water Planning for Israel Company, 232–234
technology: history, 34, 40, 45–46; green, 7, 137; impacts, 68; optimism, 7, 292; transfer, 7, 49, 169
Teggart, Frederick, 31
terrorism, 132, 134, 136
Third World, 6, 10, 61–86, 155–178, 189, 297–301; origins, 42–50
Three Mile Island, 94
Thucydides, 37
Tilly, Charles, 196
topography, 45–46
topophilia, 201
Toynbee, Arnold, 35
transboundary resource disputes, 223–245
Transvaal, 71
Tripura, 73–74
Tropical Forestry Action Plan, 10
tropics, 35, 46–50
trypanosomosiasis, 48
tsetse fly, 48–49
Turkey, 207

Uganda, 213
Ullman, Richard, 1, 129–130, 132, 188
Union Carbide, 136
United Nations, 72, 80, 81, 132, 139, 171; Conference on Environment and Development, 3, 92, 95, 111, 160, 165, 167; UNDP, 4, 156–157; UNESCO, 4; UNEP, 4, 96
United States Department of Defense, 109–124, 189, 278, 283; Army Corps of Engineers, 120–121; European Command, 118; Strategic Defense Initiative, 277; Strategic Environment Initiative, 277; Strategic Environmental Research Program, 269–288
United States government: Department of Energy, 119, 278; Environmental Protection Agency, 119–121; Fish and Wildlife Service, 120; Geologic Survey, 120, 281; Global Climate Research Program, 276; National Aeronautics and Space Administration, 276, 283; National Oceanic and Atmospheric Administration, 119–120, 276–277, 283; National Park Service, 120; National Reconnaissance Office, 273, 275; National Security Agency, 274; National Space Council, 283; Office of Technology Assessment, 283
United States foreign policy, 130, 141–142, 157–163

Vertzberger, Yugo, 143
Vest, Gary, 125n.8
Vietnam War, 136, 191
voluntarism, 32
von Ranke, Leopold, 31

Waever, Ole, 150n.32, 183n.54, 184n.61
Walt, Stephen, 52n.4, 135–137, 137
Waltz, Kenneth, 31
Wapner, Paul, 21n.12
Ward, Lester, 41
water: conflicts, 71, 206; ground, 229–232
Weber, Max, 45
Weinberg, Alvin, 206
Wells, H. G., 41–42
West Bank, 223–245
West Bengal, 73–74
Westing, Arthur, 70, 144, 151n.47, 188, 202
Woodrow Wilson Center, 156, 294
World Bank, 4, 260
World Development Report, 159
World Health Organization (WHO), 4, 96, 133

World Resources Institute (WRI), 74
World War II, 70, 136, 205, 208
Worldwatch Institute, 188
Worner, Manfred, 161–162, 172

Xiaoping, Deng, 253, 255, 258, 263,
    264

yellow fever, 48

Young, Oran, 298

Zacker, Mark, 294
Zedung, Mao, 248–249, 253–254,
    257, 263
Zemin, Jiang, 254
Zionism, 231
Ziyang, Zhao, 255